ANALYSIS OF PRETEST-POSTTEST DESIGNS

ANALYSIS OF PRETEST-POSTTEST DESIGNS

PETER L. BONATE

CRC Press
Taylor & Francis Group
Boca Raton London New York

CRC Press is an imprint of the
Taylor & Francis Group, an **informa** business
A CHAPMAN & HALL BOOK

Chapman and Hall/CRC
Taylor & Francis Group
6000 Broken Sound Parkway NW, Suite 300
Boca Raton, FL 33487-2742

First issued in paperback 2020

© 2000 by Taylor & Francis Group, LLC
CRC Press is an imprint of Taylor & Francis Group, an Informa business

No claim to original U.S. Government works

ISBN 13: 978-0-367-57898-5 (pbk)
ISBN 13: 978-1-58488-173-5 (hbk)

**Visit the Taylor & Francis Web site at
http://www.taylorandfrancis.com**

**and the CRC Press Web site at
http://www.crcpress.com**

Library of Congress Cataloging-in-Publication Data

Bonate, Peter L.
    Analysis of pretest-posttest designs / Peter L. Bonate.
        p. cm.
    Includes bibliographical references and index.
    ISBN 1-58488-173-9 (alk. paper)
    1. Experimental design. I. Title.
QA279 .B66 2000
001.4'34—dc21   00-027509
                                                                            CIP

Library of Congress Card Number 00-027509

DEDICATION

This book is dedicated to my wife, Diana, for her patience and understanding during all those Saturday and Sunday mornings I spent working on this.

ABOUT THE AUTHOR

Peter Bonate is the associate director of the pharmacometrics group at Quintiles Transnational Corporation in Kansas City, MO, in the clinical pharmacology department. He received a Ph.D. in medical neurobiology from Indiana University, an M.S. in pharmacology and toxicology from Washington State University, and an M.S. in statistics from the University of Idaho. He has more than 25 publications in the areas of pharmacokinetics, drug metabolism, bioanalytical chemistry, and drug addiction. His research interests include simulation of clinical trials, population pharmacokinetics, and drug development. He is a member of the American Association of Pharmaceutical Scientists and the American College of Clinical Pharmacology.

PREFACE

I started writing this book a few years ago when my wife was working on her doctoral dissertation. Her doctoral research involved a pretest-posttest experimental design wherein subjects were given a psychological test as a baseline, randomized to treatment groups, and asked to return a week later. Each subject was then administered one of three treatments and tested again to determine whether a treatment effect occurred. The design seemed simple enough, except that the further we looked into how to analyze the data the more confused we became. Initially, I thought to use percent change scores as a simplification and then test for differences among groups using analysis of variance (ANOVA). However, some of her professors thought we shouldn't use percent change, but rather difference scores. Others told us to avoid ANOVA altogether and use analysis of covariance. There was no consensus. Looking in the literature was no help either. The papers on analysis for this type of problem were scattered in the biomedical, educational, and psychological literature going back 30 years. It was all very confusing. I decided then that what was needed was a good review paper on how to analyze pretest-posttest data. Once I got started I realized that I had a lot more material than a review article and that a book was in order.

My goals were quite simple – to have one place for most, if not all, of the reference material which dealt with analysis of pretest-posttest designs and for that material to be useful to the average researcher. There is a large body of work, largely academic in nature, in the statistical literature which is beyond the mathematical and statistical capabilities of researchers. I consider myself relatively "mathletic" and if I couldn't understand a paper after reading it once or twice or if it couldn't be easily implemented using readily accessible computer software, it wouldn't be included in the book. I think I have succeeded for the most part. It is expected that the reader will have taken a course in basic statistics and experimental design and have at least a basic knowledge of SAS programming. I have tried to leave the derivations to a minimum and include as many real-life examples as possible from as many areas as possible. My hope is that researchers can look upon this book as a useful reference tool in dealing with problems and analyses of pretest-posttest data.

I would like to thank Chuck Davis at the University of Iowa for his help and encouragement. I would also like to thank Michele Berman of CRC Press/Chapman & Hall for her editorial assistance.

CONTENTS

CHAPTER 1

INTRODUCTION

Clinical Applications of Pretest-Posttest Data

Pretest-posttest designs fall under the broad category of paired data analysis. Paired data arise when the same experimental unit, such as a person or laboratory animal, is measured on some variable on two different occasions or at the same time under different testing conditions. There are two general types of pretest-posttest designs. The first type is when a single subject is measured on two separate occasions and the researcher wishes to determine if there is a difference between the first and second measurements. The first measurement is called the pretest, or baseline measurement, and the second measurement is called the posttest measurement. Consider the following example:

- A researcher develops a new pencil and paper test for quantifying a certain personality trait in individuals. To determine the reliability of the instrument, i.e., does the test produce the same score on different occasions for an individual, the researcher administers the test to a group of subjects once and then again a week later. The scores on both occasions were measured using the same instrument and compared against each other.

This type of study is uncontrolled because there is no control group to use for comparison. If there was a difference between pretest and posttest scores, it would be impossible to determine whether the change was due to unreliability of the measuring instrument or an actual change in the individual. These studies assume that any difference between measurements is due to unreliability of the measuring device.

The second type of pretest-posttest design is when subjects receive a treatment intervention prior to the measurement of the posttest, after completion of the pretest. Examples include:

- In a clinical trial, schizophrenic patients are randomized to receive one of two treatments: haloperidol or a new experimental drug. The baseline severity of each patient is assessed and then each subject receives one of the two medications for 6 weeks. At the end of the study each subject is assessed for the severity of their illness. In this case the treatment intervention was the drug being administered. This is a controlled study in which the treatment of interest is compared to a treatment with known effects. Note that the control group can receive standard treatment or placebo.

- A researcher is interested in determining whether endorphin levels are higher after a runner completes a marathon. Prior to beginning the race, a blood sample is taken from each subject. Immediately after the race, each runner has another blood sample collected. The blood endorphin

1

level is measured in each sample and compared within subjects. In this case, the treatment intervention was the marathon. This is an uncontrolled study.

What definitively characterizes pretest-posttest data is that two measurements are made on the same experimental unit, one measurement possibly being made prior to administration of a treatment intervention (but not necessarily so) and a temporal distance separates the collection of the posttest measurement from the pretest measurement.

A further extension of the pretest-posttest design is when multiple measurements are made after collection of the baseline or pretest variable. For example:

- A cardiologist measures the blood pressure in a group of patients longitudinally over the course of 45 years. The investigator is interested in determining if significant blood pressure changes occur during the course of a person's lifetime. This extension of the first type of pretest-posttest study is uncontrolled and typical of many prospective longitudinal studies.

- In a clinical trial, patients with high cholesterol are randomized to receive one of three treatments: placebo, a new experimental drug, and the competitor's drug. Each subject's baseline cholesterol is assessed and afterward he receives one of the three drugs for 6 months. Each subject's cholesterol is measured monthly until the conclusion of the study. This study design is more commonly called a repeated measures experimental design with baseline measurement and it is controlled.

In each example presented previously, the question of interest was either was there a difference among groups, or was there change in an individual over time. For purposes of this book we will focus on the first question, is there a difference among groups. It will be assumed that the groups differ in one or more experimental manipulation(s), or treatment intervention(s), having been applied to them. This book will not examine those situations in which each subject acts as his own control. Two such examples are:

- Suppose subjects are tested with two formulations of the same dermatological patch and the researcher wishes to determine if there is a difference in skin rash between formulations. The patches may be put on at the same time on different parts of the body (on both arms, for example) and the subject followed for some period of time. Then, at the end of the study, the degree of skin rash is measured.

- A researcher wishes to determine if a new laser procedure to correct myopia is better than traditional methods. In this study each patient may be tested with each surgical procedure in each eye. In both cases there is paired data within each patient and, in effect, each patient acts as his own control.

Another experimental design which will not be dealt with are those designs in which the data are formed in such a way so as to present each measurement at the conclusion of the study with an appropriate matched control. Here is an example:

- Suppose 20 patients are matched in such a way that each patient is matched to another person with similar characteristics of interest so that the total sample size is 40. The patients are separated into two groups such that the matched patients are each in a different group. A treatment intervention is applied, and all subjects are tested for some variable at the conclusion of the study. This type of design might be used in a case control study. Again, the data from each matched subject may be treated as paired observations.

Analysis of these data types is not easily amenable to pretest-posttest statistical techniques because pretest-posttest data implicitly assume that the value of the posttest measurement is in some manner dependent on the value of the pretest measurement, and the subsequent statistical analysis takes this dependency into account. At issue in the matched pairs design, or case-control study, is the problem of what is the pretest measurement and what is the posttest measurement since both measurements are made at the same time. If that issue can be solved, it is possible the methods that will be presented may be amenable for this type of analysis. For instance, the researcher may deem the control subject, the pretest, and the active treatment subject, the posttest. In this case the methods to be presented in this text may be used to determine if the active treatment was more effective than the control treatment.

Why Use the Pretest Data?

The goal of most experimental designs is to determine if there is a difference among groups with regard to some variable of interest after imposition of a treatment intervention. The most common method to determine if a difference exists between groups is to use Student's t-test or analysis of variance (depending on the number of groups involved) on the variable of interest after the treatment has been applied. However, this approach may not be optimal to analyze pretest-posttest data (or any type of data for that matter) because there is no guarantee that the groups are comparable at baseline. Both the analysis of variance and t-test implicitly assume that groups are comparable at baseline prior to imposition of a treatment intervention. Also, analysis of the final measurements ignores within subject variation which, if included in the analysis, may increase the researcher's ability to detect a significant difference between groups.

As an example, consider the schizophrenia study presented previously wherein each subject was randomized to one of two treatments. After 6 weeks of treatment, each subject's severity of schizophrenia was rated on a scale ranging from 0 to 100, with 0 being no observable symptoms and 100 being full-blown psychosis. The average final score in the haloperidol-treated group (the control group) was 40 ± 0.25 (mean $\pm$ standard error), whereas the score in the experimental drug treatment group was 35 ± 0.35. Using Student's 2-tailed t-test the researcher determines the p-value to be less than 0.001 and therefore concludes that the new experimental drug is more effective than traditional pharmacotherapy. In this instance, the researcher assumed that each group had comparable baseline features because each subject was randomized to each treatment. What if the baseline score in the haloperidol and experimental drug treatment group, despite randomization, was 45 ± 0.5 and 30 ± 0.6, respectively? Over the course of the treatment the haloperidol-treated group decreased 5 points, whereas the experimental drug treatment group increased 5 points. This would suggest that the experimental drug did not, in fact, have a positive effect, but rather a negative effect – an outcome completely different from the one based solely on analysis of the final data. In this case ignoring baseline variability led to the wrong conclusion.

One of the first steps in a clinical trial is the allocation of patients to treatment groups. The most important principle in subject allocation is that subjects are randomized into treatment groups without bias, but not necessarily leading to groups that have similar baseline (Altman and Dore, 1990). This principle was seen with the schizophrenia example; subjects were randomly allocated to treatments, but this alone did not insure baseline comparability between groups. Often researchers will check using null hypothesis testing whether the treatments groups have the same baseline values on average prior to or at randomization. Testing for balance is sometimes called baseline balance. In particular, the null and alternative hypotheses may be written as

$$H_o: \ \mu_1 = \mu_2 = ...\mu_k$$
$$H_a: \text{ at least one } \mu_i \neq \mu_j, \ i \neq j$$

where μ_i is the baseline value of interest for the ith group and k is the number of groups. The researcher may then use analysis of variance or some modification thereof to test for equality of group means.

Is it important for all the groups to have the same baseline values at or after randomization into treatment groups? That depends on who you ask. Many researchers and some regulatory agencies feel that baseline comparability between groups is necessary for valid clinical trials. This view is predicated on the use of statistical tests which require baseline comparability for their results to be valid. Altman and Dore (1990) analyzed 80 randomized clinical trials in 4 leading medical journals (*Annals of Internal Medicine*, *The Lancet*, *New England Journal of Medicine*, and *British Medical Journal*) and found that 58% of the studies used hypothesis tests to examine baseline comparability. Within those studies that did hypothesis testing, a total of 600 tests were done. With multiple hypothesis testing the probability of committing a Type I error (rejecting the null hypothesis when in fact it is true) is

$$p(\text{Type I error}) = 1 - (1 - \alpha)^k \qquad (1.1)$$

where k is the number of comparisons. Thus with virtual certainty many of these studies rejected the null hypothesis when in fact it was true. Herein lies the problem with baseline testing (or using any statistical test for that matter). If a test is deemed significant at the 5% level, then the very act of hypothesis testing assures that 5% of the tests done will be rejected, when in fact they may be true. This is one argument that is made against testing for baseline balance; that it inflates the Type I error rate of a study. A more persuasive argument is that supposing a study is done and it is found after the fact that baseline comparability among groups did not exist. Does that mean that the results of the study are invalid? Certainly not; it simply means that greater care must be exercised in the interpretation of the result.

Enas et al. (1990) and Senn (1994) argue that baseline homogeneity prior to or at randomization is a moot point because treatment groups can never be shown to be identical. This is one of the main reasons for randomization in the first place. Randomization attempts to balance groups with regards to known and unknown covariates and results in a controlled clinical trial. As Senn (1994) points out "balance has nothing to do with the validity of statistical inference; it is neither necessary nor sufficient for it. Balance concerns the efficiency of statistical inference..."

In other words, an unbalanced study does not invalidate the statistical analysis but does affect the ability to make conclusions regarding the results. In the schizophrenia example previously mentioned, there was a significant difference between groups, regardless of the baseline scores. Non-comparability at baseline did not invalidate this statistical test. It did, however, bring into doubt which treatment was superior. If one assumes that baseline homogeneity is not a necessary criterion for the validity of the results of a study, then statistical methods must be available to analyze data that are incomparable at baseline. Those methods that are used to analyze pretest-posttest data take into account the possible heterogeneity of baseline scores between groups and will be highly useful in this regard.

Another reason why incorporation of pretest data into a statistical analysis is important is that analysis of posttest scores alone, even when the baseline scores among groups are comparable, may not be the most powerful method to analyze the data. Statistical power is the probability of detecting a significant difference between treatments given that there really is a difference between treatments. Surely a researcher would want to maximize the statistical power of an experimental design because designs that have low statistical power may not be able to answer the central question(s) a researcher has regarding the effectiveness of some therapy. As an example, consider a study done by Yuan et al. (1997). One of the side effects of opioid medication, such as codeine or morphine, is constipation indicated by an increase in the oral-cecal transit time, i.e., the time it takes for material to move from the mouth, through the gastrointestinal tract, and out to feces. Opioid antagonists, such as naltrexone, inhibit this effect. However, naltrexone has many side effects because it can cross the blood brain barrier and interact with receptors in the brain. Methylnaltrexone, which does not cross the blood brain barrier and does not prevent analgesia, was being studied under the Orphan Drug Act as an acceptable treatment for opioid-induced constipation. In their study, 14 subjects were recruited and their baseline oral-cecal transit times were measured under placebo conditions. In the next two sessions, subjects were randomized to receive either intravenous morphine (0.05 mg/kg) or intravenous morphine and oral methylnaltrexone (19.2 mg/kg). All subjects received all three treatments. The data are presented in the top of Figure 1.1.

It is clear that in each subject morphine increased the oral-cecal transit time and that methylnaltrexone prevented it. Now consider the situation if each of the treatments were based on 3 separate groups of 14 subjects (the bottom figure of Figure 1.1). In this case it appears as though neither morphine nor the morphine and methylnaltrexone combination had any effect. There was simply not enough power to detect a difference among groups. It is evident that although there was large variation between subjects, there was much less variation within subjects. This example demonstrates the power of collecting baseline information as part of a study. By comparing treatment effects within an individual, as opposed to among individuals, there is greater evidence that the treatment was effective. This is what makes pretest-posttest data so useful, each subject can act as their own control making it easier to detect a significant treatment effect.

Cohen (1988) has recommended that a power of 0.80 should be the minimal standard in any setting, i.e., that there is an 80% chance of detecting a difference between treatments given that there is a real difference between the treatments. There are four elements which are used to define the statistical power of a test (Lipsey, 1990). These are: the statistical test itself, the alpha (α) level of significance (or probability of rejecting the null hypothesis due to chance alone), the sample size, and the effect size. The first three elements are self-explanatory; the last element may need some clarification. Consider an experiment where the researcher wishes to compare the mean values of two different groups, a treatment and a reference group. Cohen (1988) defined the effect size as the difference between treatments in units of standard deviation,

$$ES = \frac{\mu_t - \mu_r}{\sigma} \qquad (1.2)$$

where μ_t and μ_r are the respective means for the treatment and reference group and σ is the common standard deviation ($\sigma_t = \sigma_r = \sigma$). An estimate of the effect size can be made using the sample estimates of the mean and pooled standard deviation

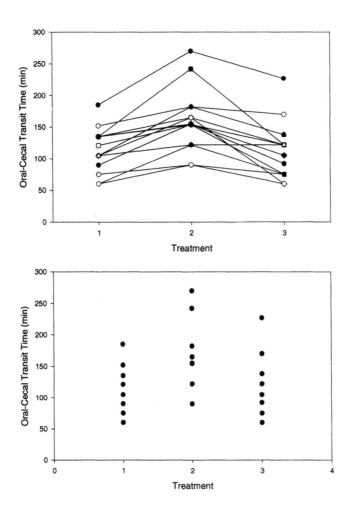

Figure 1.1: The top plot shows individual oral-cecal transit times in 14 subjects administered (1) intravenous placebo and oral placebo, (2) intravenous morphine and oral placebo, or (3) intravenous morphine and oral methylnaltrexone. The bottom plot shows what the data would look like assuming every subject was different among the groups. Data redrawn from Yuan, C.-S., Foss, J.F., Osinski, J., Toledano, A., Roizen, M.F., and Moss, J., The safety and efficacy of oral methylnaltrexone in preventing morphine-induced delay in oral cecal-transit time, *Clinical Pharmacology and Therapeutics*, 61, 467, 1997. With permission.

$$ES = \frac{\overline{X}_t - \overline{X}_r}{s} \qquad (1.3)$$

where

$$s = \sqrt{\frac{SS_t + SS_r}{v_t + v_r}} \qquad (1.4)$$

SS_t and SS_r are the sum of squares for the test and reference group, respectively, and v_t and v_r are the degrees of freedom for the test and reference group, respectively. Shown in Figure 1.2 is the difference between two means when the effect size is 1.7 or the difference between the treatment and reference means is 1.7 times their common standard deviation. As the difference between the means becomes larger, the power becomes larger as well. Figure 1.3 shows Figure 1.2 expressed in terms of power using a two-tailed t-test. Given that the sample size was 8 per group, the power was calculated to be 0.88 when $\alpha = 0.05$.

Assuming that the test statistic is fixed, what things may be manipulated to increase statistical power? One obvious solution is to increase the sample size. When the sample size is increased, the pooled standard deviation decreases, thereby increasing the effect size. This however may not be a practical solution because of monetary restrictions or difficulty in recruiting subjects with particular characteristics. The researcher may decrease the probability of committing a Type I error, i.e., α. As seen in Figure 1.3, when α is decreased, $1-\beta$, or power, increases. Again, this may not be an option because a researcher may want to feel confident that a Type I error is not being committed.

All other things being equal, the larger the effect size, the greater the power of statistical test. Thus another alternative is to increase the effect size between groups. This is often very difficult to achieve in practice. As Overall and Ashby (1991) point out "individual differences among subjects within treatment groups represent a major source of variance that must be controlled."

If during the course of an experiment we could control for differences among subjects, then we may be able to reduce the experimental variation between groups and increase the effect size. In most studies, variation within subjects is less than variation between subjects. One method to control for differences among subjects is to measure each subject's response prior to randomization and use that within subject information to control for between subject differences. Thus a basal or baseline measurement of the dependent variable of interest must be made prior to imposition of the treatment effect.

However, because we are now using within subject information in the statistical analysis, the effect size should be corrected for by the strength of the correlation between within subject measurements. We can now modify our

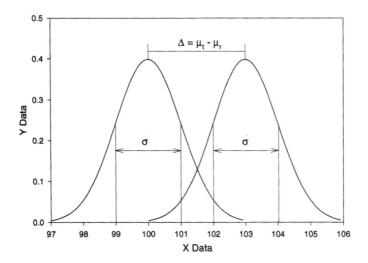

Figure 1.2: Plot showing the difference between two independent group means, Δ, expressed as an effect size difference of 1.7. Legend: μ_r, mean of reference group; μ_t, mean of test group; σ, common standard deviation.

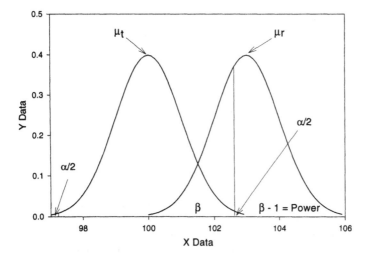

Figure 1.3: Plot showing the elements of power analysis in comparing the mean difference between two groups, assuming a 2-tailed t-test. Legend: α, the probability of rejecting the null hypothesis given that it is true; β, the probability of not rejecting the null hypothesis given that it is true.

experimental design slightly such that instead of measuring two different groups, the researcher measures each subject twice. The first measurement is before imposition of the treatment and acts as the control group. The second measurement is after imposition of the treatment and acts as the treatment group. Cohen (1988) has shown that the apparent effect size for paired data is then

$$ES = \frac{\mu_t - \mu_r}{\sigma\sqrt{1-\rho}} \qquad (1.5)$$

where ρ is the correlation between measurements made on the same subject (as will be shown in the next chapter, ρ also refers to the test-retest reliability). Figure 1.4 shows how the effect size (ES) is altered as a function of the correlation. As can be readily seen, when the correlation is greater than 0.9, the increase in power is enormous. Even still, when the correlation between measurements is moderate ($0.5 < \rho < 0.7$) the increase in effect size can be on the order of 40 to 80% (bottom of Figure 1.4). In psychological research the average correlation between measurements within an individual averages about 0.6. In medical research, the correlation may be higher. By incorporating baseline information into an analysis a researcher can significantly increase the probability of detecting a difference among groups.

In summary, including pretest measurements in the statistical analysis of the posttest measurements accounts for differences between subjects which may be present at baseline prior to randomization into treatment groups. By incorporating multiple measurements within an individual in a statistical analysis, the power of the statistical test may be increased, sometimes greatly.

Graphical Presentation of Pretest-Posttest Data

Many of the examples in this book will be presented in tabular form, primarily as a means to present the raw data for the reader to analyze on his own should he so choose. In practice, however, a table may be impractical in the presentation of data because with large sample sizes the size of the table may become unwieldy. For this reason graphical display of the data is more convenient. Graphical examination and presentation of the data is often one of the most important steps in any statistical analysis because relationships between the data, as well as the distribution of the data, can be easily visualized and presented to others. With pretest-posttest data, it is important to understand the marginal frequency distributions of the pretest and posttest data and, sometimes, the distribution of any transformed variable, such as difference scores. Much of the discussion and graphs that follow are based on McNeil (1992) and Tufte (1983).

The first step in any statistical analysis will be examination of the marginal distributions and a cursory analysis of the univariate summary statistics. Figure 1.5 presents the frequency distribution for the pretest (top)

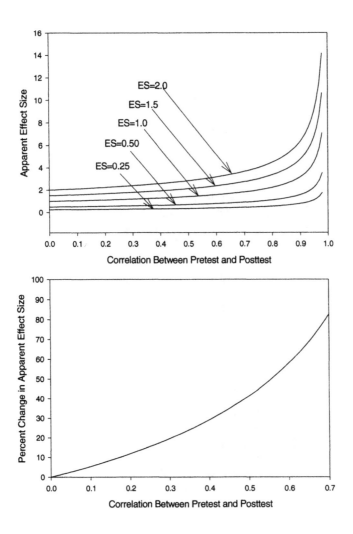

Figure 1.4: Plots showing apparent effect size as a function of the correlation between pretest and posttest (top) and percent change in effect size as a function of correlation between pretest and posttest (bottom). Data were generated using Eq. (1.5). ES is the effect size in the absence of within subject variation and was treated as fixed value. Incorporating correlation between pretest and posttest into an analysis may dramatically improve its statistical power.

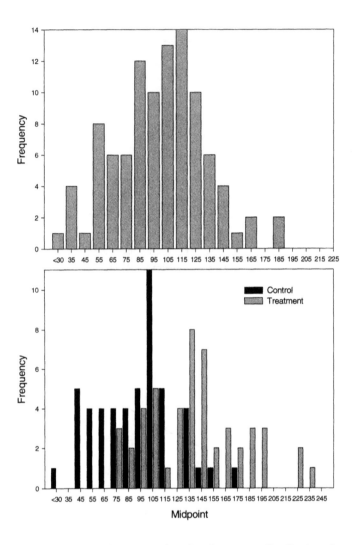

Figure 1.5: Representative example of a frequency distribution for pretest (top) and posttest (bottom) scores in a two group experimental design (treatment and control group) with a sample size of 50 in each group.

INTRODUCTION

13

and posttest scores (bottom) in a two group experimental design (one treatment and control group) with a sample size of 50 in each group. The pretest and posttest scores appear to be normally distributed. It can also be seen that the posttest scores in the treatment group tend to be higher than the posttest scores in the control group suggesting the treatment had a positive effect. The difference between the means was 48, also indicative of a difference between pretest and posttest scores.

Figure 1.6 plots as a scatter plot the relationship between pretest and posttest scores sorted by group. The control data are shown as solid circles, while the treatment group is shown as open circles. Obviously, for the expected increase in power with pretest-posttest data to be effective, the correlation between pretest and posttest must be greater than 0. In both cases, the correlation was greater than 0.6 ($p < 0.0001$), which was highly significant. Incorporation of the pretest scores into the analysis should increase our ability to detect a difference between groups. Also, the shift in means between the control and treatment groups seen in the frequency distributions appears in the scatter plot with the intercept of the treatment group being greater than the control group.

Although the majority of data in this book will be presented using histograms and scatter plots, there are numerous other graphical plots that are useful for presenting pretest-posttest data. One such plot is the tilted-line segment plot (Figure 1.7). The advantage of this type of plot is that one can readily see the relationship between pretest and posttest within each subject. Although most subject scores improved, some declined between measurement of the pretest and posttest. This plot may be useful for outlier detection, although it may become distracting when a large number of subjects is plotted. Another useful plot is the sliding-square plot and modified sliding-square plot proposed by Rosenbaum (1989). The base plot is a scatter plot of the variables of interest which is modified by the inclusion of two whisker plots of the marginal distribution of the dependent and independent variables. In a whisker plot, the center line shows the median of the data and the upper and lower line of the box show the 75% and 25% quartiles, respectively. Extreme individuals are shown individually as symbols outside the box. An example of this is shown in Figure 1.8. The problem with a plot of this type is that there are few graphics software packages that will make this plot (STATISTICA®, StatSoft Inc., Tulsa, OK is the only one that comes readily to mind).

Rosenbaum (1989) modified the sliding-square plot by adding a third whisker plot of the distribution of differences that is added at a 45% angle to the marginals. This type of plot is even more difficult to make due to the difficulty in creating plots on an angle, and few graphics software packages use will not be demonstrated. However, if more researchers begin using this type of plot, software creators may be more inclined to incorporate this as one of their basic plot types. These are only a few of the plots that can be used

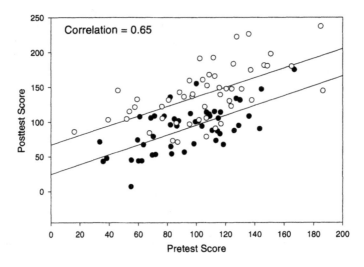

Figure 1.6: Representative example of a scatter plot of pretest and posttest scores from Figure 1.5. Legend: solid circles, control group; open circles, treatment group. Solid lines are least-squares fit to data.

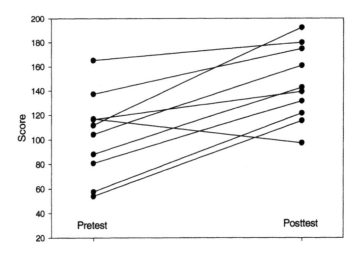

Figure 1.7: Representative tilted line-segment plot. Each circle in the pretest and posttest subgroups represents one individual.

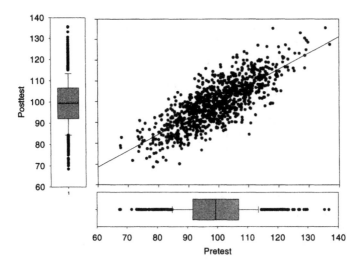

Figure 1.8: Representative sliding square plot proposed by Rosenbaum (1989). The box and whisker plots show the quartiles (25th, 50th, and 75th) for the marginal distribution of the pretest and posttest scores, respectively. The mean is the dashed line. The whiskers of the box show the 10th and 90th percentiles of the data. The scatter plot highlights the relationship between variables.

in presentations and publications. There are a variety of different ways to express data graphically and it is largely up to the imagination of the researcher on how to do so. The reader is encouraged to read McNeil (1992) for more examples.

How to Analyze Pretest-Posttest Data: Possible Solutions

Experimental designs that include the measurement of a baseline variable are often employed in behavioral studies. Despite the ubiquitous nature of such studies, how to handle data that involves baseline variables is rarely resolved in statistical books that deal with the analysis of biological or psychological data. Kaiser (1989) reviewed 50 such books and found that only 1 of them briefly mentioned the problem and the remaining 49 did not deal with the problem at all. A review of the journal literature is only partially helpful because of the myriad of statistical methods available, scattered across many different journals and other references. In addition, there is considerable debate among researchers on how to deal with baseline information, which may lead to conflicting recommendations among researchers.

Most of the methods found in the literature are based on five techniques. They are:

1.   Analysis of variance on final scores alone.
2.   Analysis of variance on difference scores.
3.   Analysis of variance on percent change scores.
4.   Analysis of covariance.
5.   Blocking by initial scores (stratification).

In the chapters that follow, the advantages and disadvantages of each of these techniques will be presented. We will also use repeated measures to analyze pretest-posttest data and examine how each of the methods may be improved based on resampling techniques.

A Note on SAS Notation

Some of the analyses in this book will be demonstrated using SAS (1990). All SAS statements will be presented in capital letters. Any variable names will be presented in italics. Sometimes the variable names will have greater than eight characters, which is illegal in SAS. They are presented in this manner to improve clarity. For some of the more complicated analyses, the SAS code will be presented in the Appendix.

Focus of the Book

The focus of this book will be how to analyze data where there is a baseline measurement collected prior to application of the treatment. The baseline measurement will either directly influence the values of the dependent variable or vary in some fashion with the dependent variable of interest. It will be the goal of this book to present each of the methods, their assumptions, possible pitfalls, SAS code and data sets for each example, and how to choose which method to use. The primary experimental design to be covered herein involves subjects who are randomized into treatment groups independent of their baseline scores. Although, the design when subjects are stratified into groups on the basis of their baseline scores will also be briefly covered. The overall goal is to present a broad, yet comprehensive, overview of analysis of pretest-posttest data with the intended audience being researchers, not statisticians.

Summary

- Pretest-posttest experimental designs are quite common in biology and psychology.
- The defining characteristic of a pretest-posttest design (for purposes of this book) is that a baseline or basal measurement of the variable of interest is made prior to randomization into treatment groups and application of the treatment of interest.
    - ◊   There is temporal distance between collection of the pretest and posttest measurements.

- Analysis of posttest scores alone may result in insufficient power to detect treatment effects.
- Many of the methods used to analyze pretest-posttest data control for baseline differences among groups.

CHAPTER 2

MEASUREMENT CONCEPTS

What is Validity?

We tend to take for granted that the things we are measuring are really the things we want to measure. This may sound confusing, but consider for a moment what an actual measurement is. A measurement attempts to quantify through some observable response from a measurement instrument or device an underlying, unobservable concept. Consider these examples:

- *The measurement of blood pressure using a blood pressure cuff*: True blood pressure is unobservable, but we measure it through a pressure transducer because there is a linear relationship between the value reported by the transducer and the pressure device used to calibrate the instrument.

- *The measurement of a personality trait using the Minnesota Multiphasic Personality Inventory (MMPI-II)*: Clearly, personality is an unobservable concept, but there may exist a positive relationship between personality constructs and certain items on the test.

- *Measurement of an analyte in an analytical assay*: In an analytical assay, we can never actually measure the analyte of interest. However, there may be a positive relationship between the absorption characteristics of the analyte and its concentration in the matrix of interest. Thus by monitoring a particular wavelength of the UV spectrum, the concentration of analyte may be related to the absorbance at that wavelength and by comparison to samples with known concentrations of analyte, the concentration in the unknown sample may be interpolated.

In each of these examples, it is the concept or surrogate that is measured, not the actual "thing" that we desire to measure.

In the physical and life sciences, we tend to think that the relationship between the unobservable and the observable is a strong one. But the natural question that arises is to what extent does a particular response represent the particular construct or unobservable variable we are interested in. That is the nature of validity–a measurement is valid if it measures what it is supposed to measure. It would make little sense to measure a person's cholesterol level using a ruler or quantify a person's IQ using a bathroom weight scale. These are all invalid measuring devices. But, validity is not an all-or-none proposition. It is a matter of degree, with some measuring instruments being more valid than others. The scale used in a doctor's office is probably more accurate and valid than the bathroom scale used in our home.

Validation of a measuring instrument is not done per se. What is done is the validation of the measuring instrument in relation to what it is being used for. A ruler is perfectly valid for the measure of length, but invalid for the measure of weight. By definition, a measuring device is valid if it has no systematic bias. It is beyond the scope of this book to systematically review and critique validity assessment, but a brief overview will be provided. The reader is referred to any text on psychological measurements or the book by Anastasi (1982) for a more thorough overview of the subject.

In general, there are three types of validity: criterion, construct, and content validity. If the construct of interest is not really unobservable and can be quantified in some manner other than using the measuring instrument of interest, then criterion validity (sometimes called predictive validity) is usually used to validate the instrument. Criterion validity is probably what everyone thinks of when they think of a measuring device being valid because it indicates the ability of a measurement to be predictive. A common example in the news almost every day is standardized achievement tests as a predictor of either future achievement in college or aptitude. These tests are constantly being used by educators and constantly being criticized by parents, but their validity is based on their ability to predict future performance. It is important to realize that it is solely the relationship between instrument and outcome that is important in criterion validity. In the example previously mentioned, if riding a bicycle with one's eyes closed correlated with achievement in college then this would be a valid measure.

Technically there are two types of criterion validity: concurrent and predictive. If the criterion of interest exists in the present, then this is concurrent validity, otherwise it is predictive validity. An example of concurrent validity is an abnormally elevated serum glucose concentration being a possible indicator of diabetes mellitus. The difference between concurrent and predictive validity can be summarized as "Does John Doe have diabetes (concurrent validity)?" or "Will John Doe get diabetes (predictive validity)?" In both cases, the methodology to establish the relationship is the same; it is the existence of the criterion at the time of measurement which differentiates the two.

Content validity consists of mapping the entire sample space of the outcome measures and then using a subsample to represent the entirety. Using an example taken from Carmines and Zeller (1979), a researcher may be interested in developing a spelling test to measure the reading ability of fourth grade students. The first step would be to define all words that a fourth-grade student should know. The next step would be to take a subsample of the sample space and develop a test to see if the student could spell the words in the subsample. As can be easily seen, defining the sample space is difficult, if not impossible, for most measurement instruments. How would one define the sample space for weight or the degree of a person's self-esteem? A second problem is that there is no rigorous means to assess content validity. Therefore, content validity is rarely used to validate a measuring instrument, and when used is seen primarily in psychology.

Construct validity measures the extent to which a particular measurement is compatible with the theoretical hypotheses related to the concept being measured. Biomedical and social scientists base most of their measurements on this type of validity, which, in contrast to the other types, requires the gradual accumulation of knowledge from different sources for the instrument to be valid. Construct validity involves three steps (Carmines and Zeller, 1979):

1. The theoretical relationship between the concepts must be defined.
2. The empirical relationship between the measures of the concepts must be defined.
3. The empirical data must be examined as to how it clarifies the construct validity of the measuring instrument.

For construct validity to be useful, there must be a theoretical basis for the concept of interest because otherwise it would be impossible to develop empirical tests that measure the concept. As Carmines and Zeller (1979) state "theoretically relevant and well-measured external variables are ... crucial to the assessment of construct validity." Thus a measure has construct validity if the empirical measurement is consistent with theoretical expectations.

Social scientists tend to put far more emphasis in the validation of their instruments than do biological and physical scientists, probably because their constructs are more difficult to quantify. Biomedical measurements are often based originally using construct validity but are later validated using criterion validity. For example, the first scale used to measure weight was developed using the laws of physics, specifically the law of gravity. Later scales were then validated by comparing their results to earlier validated scales (concurrent validity). Sometimes, however, biomedical science measurements are based face validity, another type of validity which is far less rigorous than the others and is often placed in the "second tier" of validations. Face validity implies that a test is valid if it appears on its face value to measure what it is supposed to measure. For example, a perfectly valid test for depression at face value would be the degree to which people make sad faces. That may seem ludicrous, but consider in the field of behavioral pharmacology a test used to screen for antidepressants. Rats are placed in a large beaker of water with no way to escape. Eventually the rats will "give up" and quit trying, at which time the test is over and the time recorded. Many antidepressants increase the time to "giving up" relative to controls. This test was developed for its face validity; it was thought that depressed people give up more easily than non-depressed individuals. That this test works is surprising, but also consider that stimulants tend to show up as false positives on the test.

In summary, for a measurement to be useful it must be valid and measure what it is supposed to measure. A measuring instrument may be valid for one type of measurement but invalid for another. The degree of validation a measuring instrument must possess depends on the construct the instrument is trying to measure and the degree of precision required for the instrument. As we shall see, validity is not the only criterion for a useful measurement; it should also have the property of reliability.

What is Reliability?

Suppose we measure a variable, such as weight or height, on a subject without error, i.e., with a valid measuring instrument. Then that subject's true score, T_i, would be denoted as

$$T_i = \mu + S_i \tag{2.1}$$

where μ is the population mean and S_i refers to the ith subject effect or the deviation of the ith subject from the population mean. Because not all true scores are equal amongst subjects, a collection of true scores will have some variation due to the presence of S. For example, if 10 men are weighed, not all 10 men will have the same true weight. Because T varies, T is called a random variable. Let us now define the expected value of a random variable, E(.), as the weighted average or mean of the random variable. One example that most are familiar with is the expected value of a sample collected from a normal distribution. In this case, the expected value is the population mean. The expected value of the sum of random variables is the sum of the individual expectations or sum of the means. If the long term average of all the subject effects cancels out, as it must for μ to be the population mean, then the expected value of the true scores is

$$E(T) = E(\mu) + E(S) = \mu \,.$$

Thus the expected value of the true score is the population mean.

The generic equation for the variance of the sum of random variables may be written as

$$\mathrm{Var}\left(\sum_{i=1}^{k} a_i X_i\right) = \sum_{i=1}^{k} a_i^2 \mathrm{Var}(X_i) + 2\sum\sum_{i<j} a_i a_j \mathrm{Cov}(X_i, X_j)\,. \tag{2.2}$$

Assuming the measurements are independent, then $\mathrm{Cov}(X_i, X_j) = 0$. The variance of true scores may then be written as

$$\mathrm{Var}(T) = \mathrm{Var}(\mu) + \mathrm{Var}(S) = \sigma_s^2 \tag{2.3}$$

because μ has no variability associated with it. σ_s^2 refers to inter-subject or between subject variation and reflects the degree of variation in true scores around μ. Thus we may write that T has mean μ and variance σ_s^2 or in mathematical shorthand, $T \sim \left(\mu, \sigma_s^2\right)$.

The situation where the true scores are known is impossible to achieve because all measurements are subject to some degree of error. Two types of errors occur during the measuring process. If a measurement consistently overestimates or underestimates the true value of a variable, then we say that systematic error is occurring. Systematic error affects the validity of an instrument. If, however, there is no systematic tendency either underestimating or overestimating the true value, then random error is occurring. In general, all measurements are a function of their true value plus a combination of error term(s),

$$X_i = T + C + R_i,$$ (2.4)

where X_i is the observed measurement for the ith subject, C is a constant reflecting the degree of systematic error and R_i is random error. The expected value of X is

$$E(X) = E(T) + E(C) + E(R).$$ (2.5)

We already know that $E(T) = \mu$. Systematic error, C, is a constant with no variation and reflects the inherent bias present in any measuring device. For example, suppose the scale used to weigh a person was off by 2 lb., and when nobody was on the scale it showed a value of 2. In this case, $C = 2$. It is assumed that systematic error remains the same for all measurements and never fluctuates. Thus $E(C) = 0$. Random error, on the other hand, does not remain constant but fluctuates with each measurement. For example, a subject's weight is found to be 190 lb. The person steps off the scale and then gets weighed again. This time the weight is 189 lb. This process is repeated a few more times and the recorded weights are: 190, 189, 190, 191, 191, and 190 lb. Obviously that person's weight did not change during the brief measurement period; the observed variation simply reflects random fluctuations in the measuring device. Although its value may change from one time to the next, over time its mean value is 0. Thus the expected value of R is 0. Referring back to Eq. (2.5), the expected value of X is then

$$E(X) = T + C = \mu + S_i + C$$ (2.6)

which indicates that the mean score is a function of the true score plus a component due to systematic error. Systematic error affects the accuracy of a measurement, i.e., how close the observed value is to the true value. This makes intuitive sense because if we could repeatedly measure some value on a subject we would expect that all the random measurement errors would eventually cancel themselves out, but that systematic error would still remain. When the expected value of a random variable does not equal its estimator, we say the estimator is biased. In this case, because systematic error is occurring, the expected value is said to be a biased estimator of the true value.

In the case where measurement error is occurring, the variance of X may then be written as

$$Var(X) = Var(T) + Var(R) + 2Cov(T,R)$$
$$= Var(\mu) + Var(S) + Var(R) + 2Cov(T,R).$$ (2.7)

Assuming that random error is independent of subject effects, then

$$Var(X) = Var(S) + Var(R) = \sigma_S^2 + \sigma_R^2$$ (2.8)

where σ_R^2 is referred to as residual variability and reflects random variation around true scores. Thus the variance of the observed measurements is the sum of the variance of true scores around the population mean (inter-subject) and the variance of random errors around the true scores (residual). It is apparent that random error affects the precision of a measurement, i.e., the variance of replicated measurements collected on the same experimental unit, because Eq. (2.8) differs from Eq. (2.3) by the additional presence of σ_R^2, which acts to inflate the observed variance compared to the true variance. Note that this model assumes that the true subject scores remain constant over the testing period. This might not always be the case. In this case, σ_R^2 needs to be decomposed even further to reflect a variance term for intra-subject variability and residual measurement error. Estimation of the sub-variance terms is difficult and beyond the scope of this book. For our purposes we will assume that σ_R^2 reflects only residual, unexplained measurement error.

It should be apparent that σ_R^2 and σ_S^2 cannot be distinguished using a random sample. For example, if the weights of 10 men were {145, 150, 162, 165, 171, 172, 185, 190, 210, 231 lb.}, we could not isolate that variation due to between subject variability and residual variability because both types of variability occur simultaneously. Nor could we determine the degree of systematic error that is occurring with each measurement. Throughout this book it will be assumed that systematic error does not occur, i.e., C = 0, and its contribution to the expected value of the observed scores will be ignored. Therefore the linear model for X will be written as

$$X = \mu + S + R = \mu + e \qquad (2.9)$$

where e is the overall deviation of X from μ and is equal to the sum of S and R. In this case, the expected value of X still equals μ, but the variance of X becomes

$$\mathrm{Var}(X) = \sigma_S^2 + \sigma_R^2 = \sigma^2 \qquad (2.10)$$

where σ^2 is total variance of X.

In its most simple form, two measurements are made on all individuals in a pretest-posttest design. The first is a baseline measurement, whereas the second is a measure of the variable after imposition of some treatment intervention, although this latter criterion is not necessarily so. Measuring both before and after imposition of a treatment allows the researcher to determine if a change has occurred in the state of the individual. However, in order for the assessment of change to be valid, certain conditions must be met (Ghiselli, Campbell, and Zeddeck, 1981). These are:

1.   Each subject possesses certain traits or stable characteristics which remain constant over time called true scores.

2. No systematic error occurs within each measurement, i.e., the measuring instrument is validated. Any error that is observed is random in nature and independent of other observations.
3. Any observed score is the sum of true scores and random error.
4. The measuring device remains stable over the time interval between measurement of the pretest and measurement of the posttest.

Given these assumptions, reliability refers to the precision between two measurements made on the same variable and does not refer to how accurate a measurement is. In general, the more precise a measurement, the smaller the associated variance with that measurement, and the greater the reliability of the measurement. Ghiselli, Campbell, and Zeddeck (1981) define reliability as "... the extent of unsystematic variation in the quantitative description of some characteristic of an individual when the same individual is measured a number of times." More simply put, reliability reflects the degree of similarity between multiple measurements collected on the same individual during a period in which the state of the individual does not change. As a crude example, if one's body temperature was repeatedly measured and the values obtained were: 98.6, 98.7, 98.6, 98.6, 98.6, and 98.5°F, it would appear that there was good reliability in the measurements.

Consider the case where each member of a population has been measured on two separate occasions such that the experimental conditions present at the time of the first collection are present at the time of the second measurement. Both measurements are subject to error and the first measurement in no manner influences the value of the second measurement. Therefore

$$X_{1i} = \mu + S_i + R_{1i} \qquad\qquad (2.11)$$

$$X_{2i} = \mu + S_i + R_{2i} \qquad\qquad (2.12)$$

where X_{1i} and X_{2i} are the first and second measurements, respectively, for the ith individual, $i = 1, 2, ...n$, μ and S_i are defined as before, and R_{1i} and R_{2i} are the random errors associated with the first and second measurements, respectively. Since the same measuring device is being used for both X_1 and X_2 and the same testing conditions apply during both measurement periods,

$$\sigma_{X1}^2 = \sigma_{X2}^2 = Var(S) + Var(R) = \sigma^2 . \qquad\qquad (2.13)$$

If we subtract the population mean from both sides of Eqs. (2.11) and (2.12), then

$$\left(X_{1i} - \mu\right) = (S_i - \mu) + R_{1i} \qquad\qquad (2.14)$$

$$\left(X_{2i} - \mu\right) = (S_i - \mu) + R_{2i} . \qquad\qquad (2.15)$$

By multiplying Eq. (2.14) by Eq. (2.15) and summing over the n measurements that comprise the sampled population we get

$$\sum_{i=1}^{n}(X_{1i}-\mu)(X_{2i}-\mu)=\sum_{i=1}^{n}(S_i-\mu)^2+\sum_{i=1}^{n}R_{1i}R_{2i}+$$

$$\sum_{i=1}^{n}R_{1i}(S_i-\mu)+\sum_{i=1}^{n}R_{2i}(S_i-\mu). \tag{2.16}$$

Assuming that the errors are random, i.e.,

$$\sum_{i=1}^{n}R_{X1}=\sum_{i=1}^{n}R_{X2}=0,$$

and both uncorrelated with each other and with the true scores, then

$$\sum_{i=1}^{n}(X_{1i}-\mu)(X_{2i}-\mu)=\sum_{i=1}^{n}(S_i-\mu)^2. \tag{2.17}$$

Dividing both sides of the equation by $n\sigma^2$ gives

$$\frac{\sum_{i=1}^{n}(X_{1i}-\mu)(X_{2i}-\mu)}{n\sigma^2}=\frac{\sum_{i=1}^{n}(S_i-\mu)^2}{n\sigma^2}. \tag{2.18}$$

Recognizing that the variance of the true scores is

$$\sigma_S^2=\frac{\sum_{i=1}^{n}(S_i-\mu)^2}{n}=\text{Var(S)}, \tag{2.19}$$

substituting into Eq. (2.18) and simplifying gives

$$G=\frac{\sum_{i=1}^{n}(S_i-\mu)^2}{n\sigma^2}, \tag{2.20}$$

which is equivalent to

$$G=\frac{\sigma_S^2}{\sigma^2}=\frac{\sigma_S^2}{\sigma_S^2+\sigma_R^2}. \tag{2.21}$$

G is defined as the test-retest reliability or reliability coefficient between two measurements. G will be used interchangeably with ρ, the test-retest correlation, throughout this book. From Eq. (2.21) it can be seen that the test-retest reliability between two measurements made on the same experimental unit is the proportion of the "true subject" or inter-subject variance that is contained in the observed total variance. G will always be positive and bounded on the interval (0,1) because the observed variance is in the denominator which will always be greater than or equal to inter-subject variance. If σ^2 is 100 and σ_S^2 is 70, then G = 0.70 with 70% of the observed variance attributed to the true score and the remaining 30% being attributed to random error. Table 2.1 lists the reliability coefficients for a variety of different measuring devices. The reliability of many measuring devices or tests may not be as high as one may think, especially for chemical assays routinely used for medical diagnosis.

Eq. (2.21) is just one method that can be used to compute the reliability coefficient; there are many other methods that can also be used to compute the reliability coefficient. One equally valid method to compute the reliability coefficient is the squared correlation between true and observed scores. However, this definition and the one derived in Eq. (2.21) assume that the variance of the true scores is known. In reality, this is not the case. A more common method to estimate the reliability or stability of measurements, one that does not depend on the variance of the true scores, is the test-retest correlation between measurements. Assuming no change in the state of the individual or the measuring device between measuring times, if the correlation between measurements made on the same individual are high, then pretest and posttest measurements will be similar and pretest scores will be predictive of posttest scores. The critical element here is the time period between measurements. Ferguson and Takane (1989) make the following remarks regarding test-retest reliability:

> When the same test is administered twice to the same group with a time interval separating the two administrations, some variation, fluctuation, or change in the ability or function measured may occur. The departure of (G) from unity may be construed to result in part from error and in part from changes in the ability or function measured. With many (measuring devices) the value of (G) will show a systematic decrease with increase in time interval separating the two administrations. When the time interval is short, memory effects may operate (resulting in a) spurious high correlation.

Ferguson and Takane (1989) make a good point in that longer time intervals between measurements result in a smaller reliability between the pretest and posttest measurements than shorter time intervals. On the other hand, much shorter time intervals may result in some type of carry-over from one measurement to the next, depending on the nature of the measuring device, which may bias the second measurement. Some authors (Shepard, 1981; Chuang-Stein and Tong, 1997) suggest that the reliability coefficient decays in an exponential manner over time,

$$G(t) = G(0) \cdot e^{-\lambda t}$$

(2.22)

where G(t) is the reliability coefficient at time t, G(0) is the reliability coefficient at time 0, and λ is the decay constant. If G(0) and λ are known or a good estimate is available, the reliability of a measurement at any time can be estimated.

TABLE 2.1

**RELIABILITY COEFFICIENTS FOR SOME TESTS REPORTED
IN THE LITERATURE**

Device	Reliability	Reference
Tidal Volume	0.96	van der Ent and Mulder, 1996
Glasgow Coma Scale	0.66	Menegazzi et al., 1993
Blood Pressure	0.76	de Mey and Erb, 1997
Electrocardiogram	0.82-0.96	Burdick et al., 1996; de Mey and Erb, 1997
Weschler Adult Intelligence Scale-Revised (median of subset scores in a normal elderly population)	0.71	Snow et al., 1989
Differential Diagnosis of Disease	0.61	Burdick et al., 1996
Echocardiogram	0.58-0.70	de Mey and Erb, 1997
Biochemical Analysis	see Table 2.3	McDonald, Mazzuca, and McCabe, 1983
Minnesota Multiphasic Personality Inventory (Standard Scales) in a Normal Population	1 day $\cong$ 0.83 1 – 2 weeks $\cong$ 0.75 > 1 year $\cong$ 0.40	Graham, 1987

Shepard (1981) suggests that if multiple measurements are made on the same individuals at different points in time and the conditions for measuring change apply, then G(0) and λ can be estimated as follows. If $G_{t,t+T}$ is the test-retest correlation coefficient between measurements at time t and time t+T (the period interval being T) then natural log transformation of Eq. (2.22) results in

$$Ln\left[G(t, t+T)\right] = \ln\left[G(0)\right] - \lambda T. \tag{2.23}$$

G(0) and λ can be estimated by linear regression of the natural log transformed test-retest correlation coefficients over time, where G(0) is the exponent of the intercept and λ is the slope of the line, assuming of course that the assumptions of linear regression are satisfied.

The reliability coefficient can be thought of as a measure which predicts future values. When the coefficient of reliability is low, an individual's scores on repeated measurements exhibit large variation. Conversely, when the coefficient of reliability is high, repeated measurements on an individual are nearly similar. If the reliability coefficient is equal to 1, repeated measurements of a subject results in measurements with exactly the same value and no variation. It should be stressed, however, that reliability does not bespeak accuracy. A measuring device may be very inaccurate but have high reliability. For example, an oral thermometer may give serially collected readings on the same person as: 240, 241, 240, 240, and 241°F. The reliability is near unity but clearly the accuracy is quite poor. Also, if Y = X and Z = X + C, where C is a systematic error, the reliability coefficient for X and Y is the same as for X and Z, providing $\sigma_X^2 = \sigma_Z^2$. If, however, C is a random variable and not a constant, the reliability of X and Y will be greater than the reliability for X and Z. For further in-depth discussions of reliability and how to obtain reliability coefficients for a measuring device the reader is referred to Ghiselli, Campbell, and Zeddeck (1981) or Ferguson and Takane (1989). What is important to remember is that reliability ranges from 0 to 1 and that the higher the degree of reliability the more precise measurements made on the same individual will be. Obviously a researcher should use a measuring device that has the greatest degree of reliability (all other factors being equal).

What is Regression Towards the Mean?

During the course of an experiment, measurements on many different variables may be made at many time periods on the same subjects. Regression towards the mean is important in the case of where on at least two occasions measurements involving the same variable are collected on the same subject, thereby allowing us to use the term change. The first report of regression towards the mean was by Sir Francis Galton (1885) who showed that the offspring of tall parents tended to be shorter than their parents, whereas the children of short parents tended to be taller than their parents. Galton initially termed this phenomenon regression towards mediocrity and the continued usage of the term regression stems from this initial terminology. Biologists may be more familiar with the term "law of initial values" where the "the intensity and direction of the response of any function of the organism at the moment of stimulus depends to a great extent on the initial value (level) of the function at the moment of stimulation" (Wilder, 1957). With extremely high responses there is a tendency towards no response, but with extremely low initial levels there is a tendency towards paradoxical reactions (Wilder, 1957).

Regression towards the mean (or regression to the mean) was later coined by psychologists and psychometricians in the 1960s and its usage has continued to this day. Furby (1973), Nesselroade, Stigler, and Baltes (1980), and Labouvie (1982) present comprehensive overviews of the issues and concepts involved in regression towards the mean and their main points will be presented herein. The reader is referred to the original articles for further discussion. Another excellent source of information is the June 1997 issue (Vol. 6, No. 2) of the journal *Statistical Methods in Medical Research*, edited by Stephen Senn, which is devoted in its entirely to regression towards the mean.

As an example, suppose that X (the pretest) and Y (the posttest) measure the same variable at time 1 and time 2, respectively, on the same subjects and the conditions which were present at the time of the pretest are present at the time of the posttest. Let the true scores follow a normal distribution with mean 100 and variance 100, written in mathematical shorthand as T~N(100, 100), where N refers to the normal distribution. The variance of the scores on both occasions is equal, $\sigma_{s,x}^2 = \sigma_{s,y}^2 = 100$. Also, let the variance due to random error be the same for both the pretest and posttest, $\sigma_{R1}^2 = \sigma_{R2}^2 = 25$. Thus by Eq. (2.10), $\sigma^2 = \sigma_s^2 + \sigma_r^2 = 125$. Under these conditions, 1000 normally distributed random variates were generated.

The summary statistics are presented in Table 2.2 and a scatter plot of pretest vs. posttest scores is shown in Figure 2.1. As can be seen, the calculated reliability using Eq. (2.21) was very near the theoretical reliability of 0.80. Both the squared correlation between true scores and observed scores and correlation between observed scores were very near the theoretical reliability. The test-retest reliability was also very near the theoretical reliability. The difference between the observed reliabilities and the theoretical reliabilities is due to both a combination of sample size and random error. In any case, all three methods used to calculate the reliability coefficients were very good approximations of the true value.

In Figure 2.1, which plots the scatter of all the data in from all the subjects, the paired {X,Y} data form an ellipse. Now suppose we choose any value along the X-axis and call it X'. Next, take all those subjects whose score is X' and calculate their mean score $\overline{Y}'$ at time 2. We find that for any $\{X', \overline{Y}'\}$ combination, when $\overline{X}' \neq \overline{Y}'$, the difference between $\overline{Y}'$ and $\overline{Y}$ is smaller than the difference between X' and $\overline{X}$. This situation is shown in Figure 2.2. For simplicity, the data from Figure 2.1 has been removed and replaced by an ellipse around the boundaries of the data. Also, the axes have been removed. What this plot demonstrates in lay terms is that measurements that are far from the mean on the first measurement will tend to be closer to the mean on the second measurement. This is regression towards the mean in a nutshell. More formally, Furby (1973) defined regression towards the mean as follows: given a score X' on X, the corresponding mean of Y, i.e., E(Y|X =

TABLE 2.2

**SUMMARY STATISTICS FOR 1000 NORMALLY
DISTRIBUTED, RANDOMLY GENERATED VARIATES**

Variable	Mean $(\mu = 100)$	Variance $\left(\begin{matrix} \sigma_T^2 = 100 \\ \sigma^2 = 125 \end{matrix} \right)$
True X	99.435	98.061
True Y	99.435	98.061
Observed X	99.441	124.150
Observed Y	99.479	120.323
Squared correlation between true X and observed X:		0.801
Squared correlation between true Y and observed Y:		0.788
Correlation between observed X and observed Y:		0.793
Theoretical reliability:		0.800
Reliability of X:		0.790
Reliability of Y:		0.815
Average reliability pooled across X and Y:		0.802

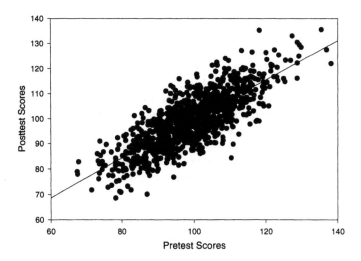

Figure 2.1: One-thousand normally distributed scores with mean 100 and variance 125. The test-retest reliability between pretest and posttest is 0.80. The solid line is the least squares linear fit to the data.

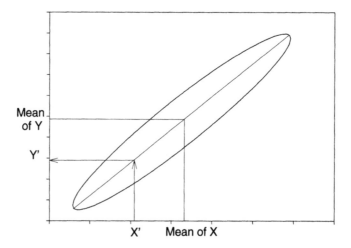

Figure 2.2: Graphical simplification of Figure 2.1 demonstrating regression towards the mean. Take any value X' and the corresponding Y' value. The difference between Y' and the mean Y value will be less than the difference between X' and the mean X value, suggesting that Y has "regressed" towards the mean.

X') is closer to E(Y) in standard deviation units than X' is to E(X) in standard deviation units.

Figure 2.3 plots the difference between posttest and pretest measurements on the same individual against his baseline pretest score. A positive difference score indicates that the posttest score was greater than the pretest score, whereas a negative difference indicates that the posttest score was less than the pretest score. It is apparent that individuals whose pretest scores were below the mean tended to have difference scores that were positive, whereas subjects whose pretest scores were above the mean tended to have difference scores that were negative. The result being an obvious negative correlation is observed between variables. This phenomenon has come to be known as regression towards the mean because subjects appear to regress towards the mean when they are measured on the same variable some later time without any type of intervention applied between measurements of the pretest and posttest. Although we have assumed equal variances between pretest and posttest, the case generalizes to real-life situations where unequal variances are present.

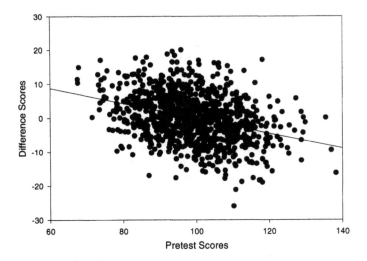

Figure 2.3: The difference between posttest and pretest scores against each subject's baseline score. Original data are in Figure 2.1. The solid line is the least squares regression and the negative slope is a hallmark of regression towards the mean. Subjects whose baseline scores are less than the mean will tend to show positive changes from baseline, whereas subjects whose baseline scores are greater than the mean will tend to show negative changes from baseline. This effect is independent of any treatment effect that might occur between measurements of the pretest and posttest.

Regression towards the mean does not occur because of some underlying biological or physical property common to the subjects being measured; it is solely a statistical phenomena and is due entirely to the properties of conditional expectation. Conditional expectation is the expectation given that some other event has already occurred. When both X and Y are normally distributed, the conditional expectation of Y, given an observed value of x is

$$E(Y|X = x) = \mu_Y + \frac{G(x - \mu_x)\sigma_Y}{\sigma_X} \qquad (2.24)$$

where G is the reliability coefficient (note that here G is calculated from the correlation between observed pretest and posttest scores upon repeated measurements of the same individual), σ_Y is the standard deviation of Y, σ_X is the standard deviation of X, μ_x is the mean of X, and μ_Y is the mean of Y. When $\mu_x = \mu_Y = \mu$ and $\sigma_x = \sigma_Y = \sigma$ then

$$E(Y|X = x) = \mu + G(x - \mu).$$ (2.25)

When $G < 1$ and $x > \mu$, $E(Y|X = x)$ will be smaller than x, whereas when $G < 1$ and $x > \mu$, $E(Y|X = x)$ will be larger than x. The net effect will be that Y is closer to the mean than x. Also, under these assumptions, the conditional variance of Y is given by

$$Var(Y|X = x) = \sigma(1 - G^2).$$ (2.26)

The variance of Y is less than the variance of X resulting in an ellipse when the data are plotted in a scatter plot, as opposed to a circle. Thus prior information, i.e., measurement of x, reduces the variance of future observations. The key to regression towards the mean is a reliability coefficient less than unity because only when $G = 1$ does $E(Y|X = x) = \mu$.

Furby (1973) points out two other interesting features of regression towards the mean. One is that X and Y can be interchanged and the regression effect still observed. Hence, for a given score on Y the corresponding score on X is closer to $\overline{X}$ than Y is to $\overline{Y}$. Thus regression towards the mean occurs whether one looks forward in time from X to Y or backwards in time from Y to X. Although this may not apply in situations where individuals are measured over time, this observation is useful in arguing the importance of regression towards the mean when non-time dependent variables are compared. Biometrician Frank Weldon wrote in 1905 [as quoted by Stigler (1997)], that

> this phenomenon of regression...is not generally understood. (V)ery few of those biologists who have tried to use (Galton's) methods have taken the trouble to understand the process by which he was led to adopt them, and we find regression spoken of as a peculiar property of living things...This view may seem plausible to those who simply consider that the mean deviation of children is less than that of their fathers; but if such persons would remember the equally obvious fact that there is also a regression of fathers on children, so that the fathers of abnormal children are on the whole less abnormal than their children, they would either have to attribute this feature of regression to a vital property by which children are able to reduce the abnormality of their parents, or else to recognize the real nature of the phenomenon they are trying to discuss.

Even today, many biologists are unfamiliar with regression towards the mean and, indeed, many biological textbooks fail to mention this topic within their pages. Another interesting feature of regression towards the mean is that X and Y need not be the same variables, but rather any two correlated variables with a less than perfect correlation. Although the development of regression towards the mean herein was predicated on pretest and posttest scores being normally distributed, regression towards the mean occurs when the marginal distributions have a distribution other than normal. In fact, any probability distribution that falls into the class of distributions called the exponential power family, of which the normal distribution is the most famous case, is subject to regression towards the mean. This includes log-normal, exponential, uniform, and Laplace distributions (Chesher, 1997).

It should be stressed that regression towards the mean always occurs with repeated measurements on the same individual. This is not some spurious phenomenon that could happen - it always happens to some degree or another. There have been a variety of reasons given for why regression towards the mean occurs but the best one is that measurements that are far removed from the mean represent relatively rare events and that the farther a value is removed from the mean the more rare that event becomes. Because rare events tend not to happen repeatedly over time, repeated measurements tend to become less rare and, consequently, closer to the mean upon repeated measurement.

Why is Regression Towards the Mean Important?

Regression towards the mean is important in measuring change because the change or difference (whether expressed as raw difference or percent change) between posttest and pretest is necessarily correlated with the pretest score. The farther a given pretest score is from its mean, the greater the amount of regression towards the mean (in absolute value), which is independent of any treatment effect that may be applied prior to measurement of Y. The posttest score on an individual will tend to be greater than his corresponding pretest score when his pretest score is below the average, whereas posttest scores will tend to decrease when pretest scores are above the average pretest score, independent of any treatment effect. This effect is very important in clinical studies where subjects are enrolled only if their baseline pretest scores are greater than or less than some value. Usually the cut-off value is far removed from the mean so that the baseline scores in those subjects accepted for the study will also be far removed from the mean. Some subjects may regress towards the mean independent of any treatments that are given to the subjects. The bottom line is that in an uncontrolled study an apparent effect may be observed with a new experimental drug that is a statistical artifact and not an actual effect. This phenomenon underlies the importance of controlled clinical studies where a placebo group is used to control for regression towards the mean.

Another example of the importance of regression towards the mean is the situation where a clinically important event is identified upon a difference from baseline. For example, some drugs prolong repolarization of the heart either intentionally (anti-arrhythmics) or unintentionally (anti-psychotics). In the case of anti-arrhythmics, the prolongation is beneficial because it stabilizes the patient's heartbeat. However, unintentional prolongation may result in sudden cardiac death, as in the case of the terfenadine and erythromycin interaction. Cardiologists have suggested that a change in the QTc interval[*] on an electrocardiogram of greater than 40 to 60 msec is clinically significant. This type of flag has been called a delta flag because it represents a change from baseline, but it completely ignores the fact that one whose baseline QTc interval is below the mean will have a tendency to have a larger delta value upon remeasurement than one whose baseline is above the mean, independent of any drug that may be present. There may certainly be other situations where regression towards the mean can impact the outcome of a study, but what is truly surprising is that most statisticians and researchers are either unaware of the phenomenon or are aware of it and choose to ignore it.

McDonald, Mazzuca, and McCabe (1983) used data in the literature to determine the expected regression effect for 15 common biochemical measurements using the known test-retest reliability of the assay used to quantify the biochemical variable. They proposed that the expected percent improvement due to regression towards the mean can be expressed as function of the deviation of a subject's pretest score from the mean and the test-retest reliability

$$\text{expected percent improvement} = \frac{Q\sigma(1-G)*100\%}{\mu + Q\sigma}, \qquad (2.27)$$

where Q is the deviation from the mean in standard deviation units. Assuming that subjects will be at least 3 standard deviation units from the mean, i.e., Q = 3, McDonald, Mazzuca, and McCabe (1983) calculated the theoretical percent improvement due solely to regression towards the mean for each of the 15 biochemical variables. These are reported in Table 2.3. As can be seen, the median of the 15 tests improved by 10% simply be remeasuring the variable on the same subject. To verify their results, the authors then examined the medical records of subjects whose initial chemistry measurements were greater than 3 standard deviations from the mean. The total number of subjects they studied ranged from 905 (magnesium) to 9340 (BUN) with most tests involving greater than 6000 subjects. They then examined those subjects whose scores were at least 3 standard deviation units from the mean and for whom posttest measurements were available. These are also reported in Table 2.3. Every one of the tests improved, i.e., became less abnormal, with a median percent change of 9.5%, a 0.5% difference from their theoretical estimates. Also, both the theoretical and observed data suggest that improvements as large as 26% could be expected to be observed solely due to regression towards the mean. The authors conclude that in many clinical trials with a placebo arm, improvements observed in the placebo group were due primarily to regression towards the mean.

---

[*] The QTc interval is one of the variables which an electrocardiogram produces and is an index of cardiac repolarization.

TABLE 2.3

THEORETICAL PERCENT IMPROVEMENT AND OBSERVED PERCENT IMPROVEMENT FOR 15 BIOCHEMICAL VARIABLES DUE TO REGRESSION TOWARDS THE MEAN

Laboratory Test	Test-Retest Reliability	Theoretical Percent Improvement	Observed Percent Improvement
Sodium	0.34	2.5	6.2
Potassium	0.38	10.6	27.1
Chloride	0.29	4.2	3.8
Carbon Dioxide	0.40	10.4	9.5
Calcium	0.37	7.8	7.0
Magnesium	0.59	8.4	20.8
Phosphorous	0.56	13.0	37.3
Total Protein	0.62	6.7	6.9
Albumin	0.56	8.8	11.3
Uric Acid	0.63	15.0	13.7
BUN	0.60	16.6	21.2
Glucose	0.57	10.0	8.7
Cholesterol	0.80	6.9	13.0
SGOT	0.47	19.0	25.6
LDH	0.51	26.0	8.9

Reprinted from McDonald, C.J., Mazzuca, S.A., and McCabe, G.P., How much of the placebo 'effect' is really statistical regression?, *Statistics in Medicine*, Copyright (1983) John Wiley & Sons, Ltd. Reproduced with permission.

Dealing with Regression Towards the Mean and How to Take Advantage of Test-Retest Reliability

Regression towards the mean is always occurring in real life. The influence it may have on the value of the posttest score, however, may be minimal and ignored during the statistical analysis. There are cases, however, when regression towards the mean may not be ignored. A researcher can use a variety of preventive and corrective measures to deal with the effect regression towards the mean has on a study. McDonald, Mazzuca, and McCabe (1983) suggest using the most reliable measurement instrument to decrease the expected variance of the posttest score. This is a commonsense suggestion and needs no further explanation. They also suggest that since single measurements are less reliable than a group of measurements, researchers should use the average of a number of different measurements of similar reliability as a measure of the baseline scores. This guideline is now common practice in clinical trial design because for many of these trials not only does the pretest serve as a covariate in the statistical analysis, it also serves as a basis for entry into a study. For example, in the Lipoprotein and Coronary Atherosclerosis study (LCAS), LDL cholesterol was determined at 2, 4, and 6 weeks prior to randomization. If the mean of the 3 determinations was 115-190 mg/dl and the absolute value of 2 of 3 possible difference scores was no more than 30 mg/dl, the patient was eligible for randomization into the study (West et al., 1996). In the LCAS, the investigators were interested in controlling for both regression towards the mean and aberrant laboratory results.

Averaging pretest scores results in two improvements to the data: minimizing regression towards the mean and improving the precision between measurements (Shepard, 1981). Suppose a subject is tested repeatedly on n different days and within each day, m within-session measurements are recorded. If G is test-retest reliability on the same subject on different testing sessions (between-visit) and ρ_r is the correlation between measurements X and Y on the same subject within the same visit, then

$$\overline{\overline{G}} = \frac{1}{1 + \frac{1}{nm}\left[\frac{1-G}{G} + \left(\frac{\rho_r}{G} - 1\right)(m-1)\right]} \qquad (2.28)$$

where $\overline{\overline{G}}$ is the reliability coefficient across an arbitrary number of visits and within-session replications. Notice that when a single measurement is made within a single testing session, i.e., n = 1 and m = 1, $\overline{\overline{G}}$ simplifies to G, the test-retest reliability defined in Eq. (2.21). Figure 2.4 plots the reliability coefficient as a function of the number of inter- and intra-subject measurements. As n, the number of between-session measurements, increases to infinity, $\overline{\overline{G}}$ approaches its limit of 1. Although $\overline{\overline{G}}$ also approaches its limit as the number of within-session measurements is increased, $\overline{\overline{G}}$ approaches its limit at a much faster rate when the number of inter-trial measurements, n, is increased. Thus better reliability is obtained over multiple inter-session measurements than over an equal number of within-session measurements.

The utility of Eq. (2.28) becomes apparent when it is realized that the desired level of reliability can be achieved either through manipulating the number of measurements within a visit, by manipulating the number of measurements between visits, or by some combination of both. As an example, consider the case reported by Shepard (1981) where a person's blood pressure was repeatedly measured. For systolic pressure in the population at large, the between-session correlation was $\rho = 0.751$ and the within-session correlation was $\rho_x = 0.860$. Usually the within-visit correlation is greater than the between subject correlation because the interval between measurements is shorter. Suppose a reliability coefficient of at least 0.85 was required. The

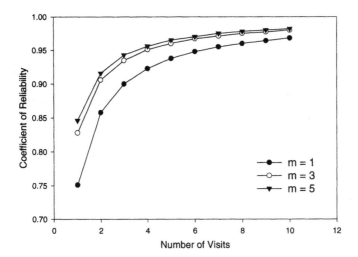

Figure 2.4: The reliability coefficient as a function of the number of intra-(m) and inter-visit (n) measurements. As either n or m increases the reliability of a measuring device increases. However, for a given number of measurements, greater reliability is achieved when more measurements are done on separate occasions than when repeatedly done on the same visit.

researcher may make three measurements within a single session or one measurement on two different sessions to achieve the desired reliability.

In some studies, only subjects with extreme pretest scores may enroll in the study. One problem with repeated pretest measurements is that they lead to increased effort, time, and, probably most important, money. It would be very useful to have a procedure to screen out those subjects that would be unlikely to enroll in a study after the first or second pretest measurement, thus saving the money and effort required to obtain the remaining pretest estimates. Suppose that the average of n values will be used as an estimate of the pretest and that this value will also be used to determine if a subject will be eligible for participation in a study. A good example of this might be a clinical trial designed to study a novel pharmacotherapy for high cholesterol. It makes little sense to enroll subjects who do not have hypercholesteremia, but have artifactually high cholesterol which returns to normal limits on repeated measurement. In the Cholesterol and Recurrent Events trial (CARE), a clinical trial assessing the efficacy of pravastatin, an HMG-CoA reductase inhibitor, which reduces fatal coronary heart disease and myocardial infarction, two LDL cholesterol measurements were made. One criterion for enrollment was that the mean of both measurements be within the interval 115 to 174 mg/dL.

In this study the researchers wanted to collect second cholesterol measurements only in those subjects whose probabilities of being eligible for the study remained high given their first cholesterol value. In other words, they wanted to be able to weed out those subjects after the first measurements whose likelihood of enrolling in the study after two measurements was small.

Assuming that n measurements have equal test-retest reliability, G, amongst any two measurements (discounting correlation between the same measurements), the probability that $\bar{X}_n$ will be an eligible average given the average of m measurement, $\bar{X}_m$, (m < n) is

$$p(e_1 \leq \bar{X}_n \leq e_2 \mid \bar{X}_m) = \Phi_z \left[\frac{e_2 - \mu_p}{\sqrt{v_p}} \right] - \Phi_z \left[\frac{e_1 - \mu_p}{\sqrt{v_p}} \right] \qquad (2.29)$$

where e_1 and e_2 are the lower and upper limits, respectively, of acceptability into the study, $\Phi(.)$ is the cumulative distribution function of the standard normal distribution,

$$\mu_p = \mu_n + \frac{m}{n} \left[\frac{1+(n-1)G}{1+(m-1)G} \right] (\bar{X}_m - \mu_m), \qquad (2.30)$$

$$v_p = \frac{1+(n-1)G}{n} \left[1 - \frac{m}{n} \left(\frac{1+(n-1)G}{1+(m-1)G} \right) \right] \sigma^2, \qquad (2.31)$$

μ_n is the population mean of n measurements, and μ_m is the population mean of m measurements (Moye et al., 1996). Eq. (2.29) can be solved iteratively to find critical values for $\bar{X}_m$ given a predefined level of probability set by the investigator. The difficulty with using Eq. (2.29) is the reliance on having to know *a priori* the population mean, the population standard deviation, and the population correlation.

As an example, using the data and example provided by Moye et al. (1996), what is the probability that an individual can enroll in the CARE trial after a single trial? Assume that the population mean and standard deviation for the first and second measurements are 137.3 ± 29.4 and 135.8 ± 29.3 mg/dL, respectively, and the test-retest correlation between measurements is 0.79. The probability that an individual will be eligible for the study is given by Figure 2.5. With almost virtual certainty, if someone has an initial LDL cholesterol of less than 90 mg/dL or greater than 210 mg/dL, they will not be eligible for the study. If the researcher wishes to be 90% confident that a subject will be eligible for the study, then taking the upper and lower 5% interval of the distribution of probabilities will suffice as cut-off values. In this instance, the upper and lower cut-off values were approximately 95 and 203 mg/dL.

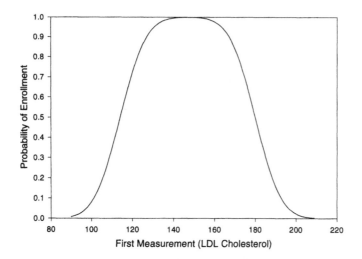

Figure 2.5: Probability of enrollment in the CARE clinical trial as a function of baseline LDL cholesterol. Reprinted from *Controlled Clinical Trials*, 17, Moye, L.A., Davis, B.R., Sacks, F., Cole, T., Brown, L., and Hawkins, C.M., Decision rules for predicting future lipid values in screening for a cholesterol reduction clinical trial, 536-546, Copyright (1996), with permission of Elsevier Science.

Davis (1976) has made a few recommendations regarding regression towards the mean in studies where individuals are selected for enrollment into a study on the basis of their pretest scores. His first suggestion was that instead of using a single measurement upon which to base patient selection, the average of a number of pretest measurements be used instead. This concurs with other recommendations. His second suggestion was that if individuals are chosen on the basis of pretest scores, second pretest measurements be used in the statistical analysis to determine whether changes have occurred. One outcome of this suggestion is that if the test-retest correlation between the first pretest and posttest measurements is the same as the test-retest correlation between the second pretest measurements and the posttest measurements, there will be no effect due to regression towards the mean in the statistical analysis. An extension of this is that when multiple pretest measurements ($n_1 + n_2 = n$) are available, n_1 measurements be used in which to base patient selection and n_2 measurements be used as the baseline upon which to determine if a change has occurred.

The methods discussed so far for dealing with regression towards the mean have been preventive in nature. What happens after a study is completed in which the preventive measures were not used and it is determined that regression towards the mean is significantly influencing the estimation of the treatment effect? Luckily there are post-hoc procedures that can be used to "correct" for the effect regression towards the mean has on the estimation of the treatment effect. These methods all involve adjusting the posttest scores by some factor and then performing the statistical analysis on so-called corrected scores. If Y is the observed posttest measurement, then the adjusted posttest measurement, Y_{adj}, is given by

$$Y_{adj} = Y - \frac{\sigma_Y}{\sigma_x}(G-1)(X-\mu) \qquad (2.32)$$

where G is the test-retest reliability coefficient, σ_Y and σ_x are the standard deviations for the posttest (s_Y) and pretest (s_x), respectively, and μ is the average pretest score (Chuang-Stein, 1993; Chaung-Stein and Tong, 1997).

The second term in Eq. (2.32), $\frac{\sigma_y}{\sigma_x}(G-1)(X-\mu)$, reflects the amount of regression towards the mean. It can be seen from Eq. (2.32) that when the posttest score is greater than the mean, the adjusted score will be less than the observed value. Conversely, when the observed posttest score is below the mean, the adjusted posttest score will be above the mean. The net effect of this is that the observed and adjusted scores will be highly correlated and the adjusted posttest scores will have the same mean as the observed posttest scores, but will have a variance slightly greater than the observed posttest scores. This is illustrated in the top of Figure 2.6, which includes plots of the adjusted posttest scores against the observed posttest scores from Figure 2.1. The line in the figure shows the line of unity. The adjusted posttest scores have a mean of 100 and a variance of 124 compared to a mean of 100 and a variance of 120 for the observed scores. The bottom figure plots the amount of regression towards the mean as a function of the observed posttest score. There is a direct linear relationship between observed posttest scores and the amount of regression towards the mean. The maximal amount of regression towards the mean was 6.5; thus, at most, 6.5% of the observed posttest scores was due to regression towards the mean, not a treatment effect. Other methods are available for correcting for regression towards the mean, but are more technically complex. The reader is referred to Lin and Hughes (1997) for a more thorough exposition of these techniques.

Another useful post-hoc procedure is to use analysis of covariance (Chapter 5). As will be seen in Chapter 5, analysis of covariance explicitly takes into account the dependence of the posttest scores on the pretest scores, as well as controlling regression towards the mean, whereas the other statistical methods which will be discussed herein do not.

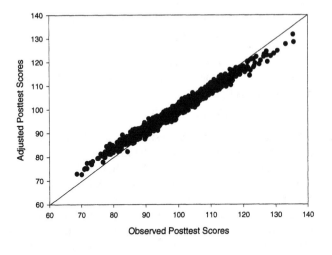

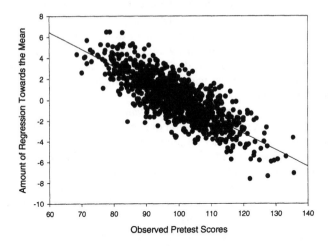

Figure 2.6: Adjusted posttest scores plotted against original posttest scores (top) and the amount of regression towards the mean against baseline scores (bottom). Adjusted posttest scores are highly correlated (solid line is the line of unity) and have the same mean as unadjusted posttest scores, but have a larger variance. The bottom plot shows that, at most, 6.5 units of the observed posttest score was due to regression towards the mean.

What is Pretest Sensitization?

One assumption to be made in future chapters is that although the pretest influences the posttest score, it does not influence the treatment effect. In other words, suppose the model for the posttest score, Y_i, is given by

$$Y_i = \mu_i + \tau + e_i \tag{2.33}$$

where μ_i is the ith subject's true score, τ is the treatment effect, and e_i is the error term. It is assumed that the covariance between the treatment effect and the true pretest score, μ_i, is 0, i.e.,

$$\mathrm{Cov}(\mu_i, \tau) = 0. \tag{2.34}$$

Thus the treatment effect and true pretest score are independent and do not influence each other. This may not always be the case. There is considerable evidence in psychological and sociological studies that when subjects have made the connection between the object of the pretest and the nature of the treatment, they may not respond with their true feelings. Subjects will tend to bias their answers to make them more palpable to either the researcher or the public at large. When this occurs we say that the pretest has *sensitized* subjects to the treatment. This occurs quite often in attitudes and opinions research and in cases where learning occurs between administration of the pretest and posttest. As Lana (1969) states "any manipulation of the subject or his environment by the experimenter prior to the advent of the experimental treatment, which is to be followed by some measure of performance, allows for the possibility that the result is due either to the effect of the treatment or to the interaction of the treatment with the prior manipulation". Most researchers tend to ignore the latter possibility.

Consider the case where a researcher wants to measure attitudes towards Judaism before and after administration of a treatment intervention designed to educate the subjects about other religious groups. In the study, subjects report to a room where they are administered questionnaires probing their feelings towards other religions. They are then randomized to either watch a video on the Holocaust or on how gold is mined. The latter treatment is designed to act as a neutral or control task. The next week subjects return and retake the questionnaire again. Clearly, in this instance, it would be difficult to hide the intention of the treatment from the subjects assigned to the Holocaust video group because of the nature of the pretest.

By definition, pretest-sensitization occurs when $\mathrm{Cov}(\mu_i, \tau) \neq 0$ but, in fact, $\tau_i^* = f(\mu_i, \tau)$, where τ_i^* is the observed treatment effect for the ith subject and f(.) indicates that the observed treatment effect is a function of the true population treatment effect and the true pretest score. Thus there is a differential effect between groups,

$$Y_{1i} = \mu_i + \tau_1 + e_{1i} \text{ for the control group}$$

and

$$Y_{2i} = \mu_i + f(\mu_i, \tau_2) + e_{2i} \text{ for the treatment group.}$$

In this instance we say that the pretest has sensitized individuals to the treatment of interest; hence the phrase pretest sensitization.

To see how an interaction between the pretest and treatment effect alters the expected value of difference scores, which will be discussed in the next chapter, suppose the pretest score is given by

$$X_i = \mu + S_i + e_i, \tag{2.35}$$

the posttest score is given by

$$Y_i = \mu + S_i + f(\mu + S_i, \tau_i) + e_i, \tag{2.36}$$

and that X and Y are continuous random variables. Suppose further that the pretest treatment interaction can be modeled by constant treatment effect and a multiplicative interaction term which is dependent on the subject's true score,

$$f(\mu, \tau_i) = \tau_i + \beta \cdot \mu_i \cdot \tau_i \tag{2.37}$$

where β is a proportionality constant and μ_i is the ith subject's true pretest score. Substituting Eq. (2.37) into Eq. (2.36) gives

$$Y_i = \mu_i + \tau_i + \beta \cdot \mu_i \cdot \tau_i + e_i. \tag{2.38}$$

In developing the expected value of difference scores, we subtracted the posttest from the pretest. Subtracting Eq. (2.35) from (2.38) and taking the expected value gives

$$E(Y\text{-}X) = \tau_i + \beta \cdot \mu_i \cdot \tau_i = \tau_i (1 + \beta\mu_i) \tag{2.39}$$

where the farthest right term in Eq. (2.39) is the sensitization effect. When $Cov(\mu_i, \tau) = 0$, the expected value of difference scores is [see Eq. (3.18) and its derivation]

$$E(Y - X) = \tau. \tag{2.40}$$

In pretest sensitization, i.e., when the treatment and covariate interact, we are left with a estimate of the treatment effect which cannot be assessed using ordinary methods.

Many authors suggest that the appropriate model for pretest sensitization is to assume that the treatment affects the posttest scores differentially, that there will be proportional change due to treatment effect, not an additive one (James, 1973; Senn and Brown, 1985, 1989). In other words, sometimes the treatment effect is not a constant across all subjects, but affects some individuals to a different extent than others. Often this differential treatment effect is dependent on the baseline of the individual. In this case the linear model for the posttest score is given by

$$Y_i = \mu + \gamma\rho(X_i - \mu) + e_i \tag{2.41}$$

where μ is the population mean pretest score, X_i is the ith subject's pretest score, and γ is the proportionality factor for treatment effect. When $\gamma = 1$, Eq.

(2.41) simplifies to the additive linear model assuming no treatment effect. When $\gamma < 1$, the treatment effect is related to the initial baseline measurement. Chen and Cox (1992) have shown that in the case where n pretest measurements are available, but only m posttest measurements are made, such that $\dfrac{n}{m} \cong 0$, then γ and ρ can be estimated by

$$\hat{\rho} = \pm\sqrt{1 - \frac{\sum_{i=1}^{n}\left[(Y_i - \mu) - a(X_i - \mu)\right]^2}{n\sigma^2}} \qquad (2.42)$$

$$\hat{\gamma} = \frac{a}{\hat{\rho}} \qquad (2.43)$$

where μ and σ are estimated from the sample estimates of the mean and standard deviation of all n pretest scores, and

$$a = \frac{\sum_{i=1}^{n}(X_i - \mu)(Y_i - \mu)}{\sum_{i=1}^{n}(X_i - \mu)^2} \qquad (2.44)$$

and $\hat{\rho}$ is the same sign as a. Note that Eq. (2.42)-(2.44) are calculated using only the first n samples, but that the estimates for μ and σ are calculated using all n+m samples.

Controlling for Pretest Sensitization with Factorial Designs

Someone once said that the best offense is a good defense. Likewise, the best method to control for pretest sensitization is to use an experimental design that isolates the interaction between pretest and posttest, i.e., to control for pretest sensitization prior to the start of the experiment. One such design is a 2^n factorial design, where n is the number of levels in the treatment group and one of the factors is a dichotomous qualitative variable "Pretest Administration" in which the possible outcomes are either "Yes" or "No." Recall that a factor is a particular class of related treatments and that a factorial design is one where the effects of a specific combination of factors are investigated simultaneously.

The most commonly seen design is a 2^2 design, called the Solomon-four in honor of the researcher who first examined this problem (Solomon, 1949). In this experimental design, four groups of subjects are used (Table 2.4). One group receives the pretest and the treatment intervention (Group 4), another group receives the treatment intervention without being given the pretest (Group 3), another group is given the pretest without being given the treatment

TABLE 2.4

BASIC LAYOUT OF A 2^2 FACTORIAL
OR SOLOMON-FOUR DESIGN

		Administer Pretest	
		No	Yes
Administer Treatment	No	Group 1	Group 2
	Yes	Group 3	Group 4

(Group 2), and the last group is given neither the pretest nor the treatment intervention (Group 1). All four groups are given the posttest, which are then analyzed using an ANOVA. A statistically significant interaction effect is evidence of pretest sensitization. The reader is referred to Kirk (1982) or Peterson (1985) for details on factorial designs.

As with any experimental design, certain assumptions must be made for the ANOVA to be valid. First, subjects are chosen at random from the population at large and then randomly assigned to one group, the assumption being that any subject in any group is similar in characteristics to any subject in another group. However, given the nature of the experimental design it is impossible to test this assumption because only half the subjects are administered the pretest. Second, the data are normally distributed with constant variance and each subject's scores are independent of the other subject's scores. This assumption is seen with many of the statistical tests which have been used up to this point.

A couple of points relating to this topic must be made. First, in this design the pretest cannot be treated as a continuous variable; it is a qualitative variable with two levels ("Yes" or "No") depending on whether or not the subject had a pretest measurement. Second, whereas in future chapters it will be advocated to include the pretest as a covariate in the linear model, this cannot be done here because only some of the subjects actually take the pretest. In fact, this is the biggest drawback of the design. Only posttest scores alone can be analyzed with any within subject information being lost in the process. Third, the Solomon-four can be expanded to include both additional treatments. For example, a 2^3 design could have two active treatment levels and a no-treatment level, a 2^4 design would have three active treatment levels and a no-treatment level, etc. Of course, the disadvantage of the factorial design is that as the number of treatment levels increases there is a geometric increase in the required number of groups and subjects, thereby possibly making the experiment prohibitive.

In factorial experimental designs, simple effects refer to the differences between the levels of the factors and main effects refer to the average of the simple effects. The difference in main effects for each level is referred to as the interaction. The presence of a positive interaction term is indicative of pretest sensitization. Table 2.5 presents the layout of simple, main, and interaction effects from a 2^2 factorial design. The first step in interpreting interaction is to plot the cell or group means and connect them with a straight line(s). Interaction is present when any portion of the lines connecting the means are not parallel. If the treatments do not interact, the lines will be parallel. Figure 2.7 demonstrates this for the data in Table 2.5. Both plots exhibit nonparallel lines between factor comparisons indicating that interaction is occurring between factors. Also, subjects that were administered the pretest had significantly higher posttest scores after having received the treatment than subjects that did not receive the treatment – the hallmark of pretest sensitization.

There is debate in the literature as to how to interpret main effects in the presence of interaction. One school of thought (the predominant one) is that if the interaction effect is significant, the main effects have no meaning, whether they are significant or not (Peterson, 1985). They argue that a main effect assesses the constant effect of a predictor variable across all levels of the other variables. Thus when interaction is present no such constant effect occurs, the treatment effect depends on a particular level of the pretest, and the main effect has little meaning when averaged over all levels of the pretest. If the interaction term is not significant, all the information is contained in the main effects, the treatment effect is independent of pretest effect, and general conclusions regarding superiority or efficacy of a treatment can be made.

Others argue that main effects are interpretable in the presence of interaction because main effects are the average of an independent variable across all other factors. Jaccard (1997) says that interaction occurs when the effect of an independent variable on a dependent variable differs depending on the value of a third variable, called the moderator variable. This requires the analyst to define what is the independent variable and what is the moderator. In the case of pretest-posttest data, the independent variable is the treatment of interest and the moderator variable is the pretest measurement because it is entirely possible for the pretest measurement to modify the impact of a treatment on a group of individuals.

According to the first school of thought, when interaction is present, every level of factor A should be compared only against every level of factor B and vice-versa. However, Jaccard (1997) recommends testing the effects of the independent variable only at each level of the moderator, thus cutting the number of hypothesis tests made in half. Although this may seem like splitting hairs, conceptually it makes a lot of sense. A researcher is not necessarily interested in the pretest factor; what is of importance is the treatment effect. In addition, the probability of making a Type I error decreases by decreasing the number of hypothesis tests being made.

It is up to the researcher to make his own decision regarding the interpretation of interaction effects. One can quite easily find examples of

TABLE 2.5

**SIMPLE, MAIN, AND INTERACTION EFFECTS IN
A 2X2 FACTORIAL DESIGN WHEN
INTERACTION IS PRESENT**

		Administer Pretest		Simple Effect	Main Effect	Interaction
		No	Yes			
Administer	No	20	22	2	13	22
Treatment	Yes	26	50	24		
Simple Effect		6	28			
Main Effect		17				
Interaction		21				

Note: The dependent variable is the posttest score. Simple effects are the difference between factor levels. Main effects are the average of the simple effects and interaction is the difference of simple effects. Pretest sensitization is reflected in dissimilar interaction terms.

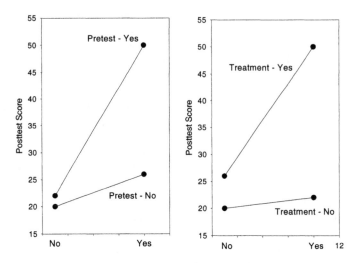

Figure 2.7: Plot demonstrating pretest sensitization using the data in Table 2.5. A differential treatment effect is observed depending on whether subjects were exposed to the pretest prior to the treatment intervention.

both schools in the literature, although in all likelihood the reason most researchers report the p-values for main effects in the presence of interaction is not because of some theory or philosophical school the researchers have ascribed to, but rather because they are either naive in the nuances of factorial designs or choose to ignore the interaction. The reader is referred to Kirk (1982), Jaccard, Turriso, and Wan (1990), or Jaccard (1997) for further details on how to handle and interpret main effects in the presence of interaction.

Alternative Methods for Controlling for Pretest Sensitization

If pretest sensitization is occurring then the researcher may wish to use a pretest that "lacks the obvious and telltale characteristics of the pretest" or increase the time interval between pretest and posttest (Lana, 1969). Of course, by increasing the time interval between pretest and posttest the researcher runs the risk of increasing the probability that some other external factors are responsible for any apparent treatment effects.

Summary

- No measurement has perfect reliability. Every measurement has some degree of error associated with it, both systematic and random in nature.
- Compounded with the inherent degree of error associated with every measurement is that regression towards the mean occurs whenever a subject is measured on at least two occasions.
- Regression towards the mean is independent of any treatment effects that may be applied between collection of the pretest and subsequent measurements.
  - ◊ When an individual's pretest score is very different from the mean pretest score, regression towards the mean becomes an important issue and may bias the estimation and/or detection of treatment effects.
  - ◊ In particular is the case where subjects are enrolled in a study based on their pretest score. In these subjects, the effect regression towards the mean will have on the estimation of the treatment effect will probably be quite substantial.
- It is often assumed that the pretest and treatment effect do not interact, but that may not necessarily be the case.
- Careful examination of the main effect plots should be used to reveal the presence of a treatment by pretest interaction.
- If a researcher believes that pretest sensitization may occur, the best method to control for it is by the appropriate experimental design.
  - ◊ If, however, an experiment suggests that pretest sensitization has occurred, it is possible to estimate the treatment effect using a modified linear model where the treatment effect is a function of the pretest score.

CHAPTER 3

DIFFERENCE SCORES

Statistical data analysis is model driven. Certain assumptions must be made regarding the distribution of the data and the dependency between the pretest and posttest measurements in order to obtain the significance of a treatment effect. We saw the beginnings of this in the last chapter where we specified the linear model for pretest and posttest measurements. This chapter and the next (Relative Change Functions) describe the most common methods used to analyze pretest-posttest data. Recall from the previous chapter that in order to assess "change," some measure of baseline activity, which we have called the pretest, must be made prior to application of the treatment. Also recall that the expected value of the posttest score, the measurement collected after application of the treatment, is dependent on the pretest score. As was seen in Chapter 1, analysis of posttest scores without considering the pretest scores may result in analysis bias and erroneous conclusions.

Both difference scores and relative change scores collapse the pretest and posttest scores into a single score thereby simplifying the problem from a multivariate one to a univariate one. Difference and relative change scores can then be analyzed using any of a myriad of univariate statistical methods. They have an advantage in that the transformed variable is easily interpreted either as a net gain or loss score (for difference scores) or as percent change from baseline (for relative change functions).

Definition and Assumptions

In order for difference scores to be used, both the pretest and posttest scores must be either a continuous random variable or interval data, thus ensuring the difference scores to be continuous or interval in nature. It is also necessary for the pretest and posttest scores to be measured using the same device or instrument. In addition, both pretest and posttest scores must have the same units, thus ensuring the difference scores to be of the same measuring units. For instance, suppose weight was measured before and after treatment. Although the same measuring device was used for collection of the pretest and posttest measurements, it would be inappropriate to compute difference scores if the posttest measurement was reported in kilograms as opposed to pounds, which were reported in the pretest measurement. It is not necessary for pretest, posttest, or difference scores to be normally distributed (although that would be nice). Also (and this applies to all the statistical methods presented in each of the chapters) the time interval between collection of the pretest and posttest measurements is the same for all subjects or at the very least, the reliability between measurements is constant for all subjects.

A difference score is defined as the raw change in score between posttest and pretest:

$$\Delta = Y - X, \tag{3.1}$$

where Δ is the difference score, Y is the posttest score, and X is the pretest or baseline score. If Δ is greater than 0, a net gain has occurred. Conversely, if Δ is less than 0, a net loss has occurred. As can be seen, difference scores offer the advantage that they are easily interpreted. For this reason, analysis of difference scores is often seen in the literature.

Case 1: The Absence of a Treatment Intervention Between Measurement of the Pretest and Posttest Scores

Consider the case where an experimenter measures some variable on n different subjects on two different occasions. The testing occasions may be generically called the pretest and posttest, respectively. The question to answer is: "was there a significant change from or difference between the two testing occasions?" The linear model for the pretest, X, can be written as

$$X = \mu + S_i + e_{i1} \tag{3.2}$$

and the linear model for the posttest, Y, can be written as

$$Y = \mu + S_i + e_{i2} \tag{3.3}$$

where μ is the population mean, S_i is the ith subject effect and e_{i1} and e_{i2} are the residual random error terms with expected value 0 and variance σ^2. Subtracting Eq. (3.2) from Eq. (3.3) and assuming that the true subject effect remains a constant, the difference scores can be expressed as

$$Y_i - X_i = \Delta = e_{i2} - e_{i1}. \tag{3.4}$$

Clearly, difference scores are not constants, but are random variables because Eq. (3.17) contains error terms. Since the expected values of the residuals are 0, the expected value of the difference scores is

$$E(Y - X) = E(\Delta) = 0. \tag{3.5}$$

The variance of difference scores can be computed using the rules of expectation. Recall that the generic equation for the variance of the sum of random variables may be written as

$$\mathrm{Var}\left(\sum_{i=1}^{k} a_i X_i\right) = \sum_{i=1}^{k} a_i^2 \mathrm{Var}(X_i) + 2\sum\sum_{i<j} a_i a_j \mathrm{Cov}(X_i, X_j). \tag{3.6}$$

Hence, the expected value of the difference score can be written as

$$\mathrm{Var}(Y - X) = \mathrm{Var}(X - Y) = \mathrm{Var}(Y) + \mathrm{Var}(X) - 2\mathrm{Cov}(X,Y). \tag{3.7}$$

However, the covariance between X and Y can be expressed as a function of the correlation coefficient between X and Y, ρ,

$$\rho = \frac{\text{Cov(X,Y)}}{\sqrt{\text{Var(X)}}\sqrt{\text{Var(Y)}}} \tag{3.8}$$

such that Eq. (3.7) can be rewritten as

$$\text{Var}(Y-X) = \text{Var}(X) + \text{Var}(Y) - 2\rho\sqrt{\text{Var}(X)}\sqrt{\text{Var}(Y)}. \tag{3.9}$$

Note that ρ is the same as the test-retest reliability between X and Y. Since pretest-posttest measures are often collected on the same variable and have the same variance, Eq. (3.9) can be simplified to

$$\text{Var}(Y-X) = \text{Var}(\Delta) = 2\text{Var}(X)[1-\rho]. \tag{3.10}$$

Thus the variance of the difference scores is a function of both the variance of the measuring device and the reliability between pretest and posttest scores. Based on the above model the variance of the difference scores will have greater precision than the sum of the variances. For this reason, an increase in precision means difference scores are analyzed instead of posttest scores.

The null hypothesis (H_o) and alternative hypothesis (H_a) for no difference between pretest and posttest scores is

$$H_o: \Delta = 0$$
$$H_a: \Delta \neq 0$$

where Δ is the population difference score. One method that can be used to test the null hypothesis is the paired samples t-test. The first step in the paired samples t-test is to compute the difference score, Δ, for each subject using Eq. (3.1) and then compute the variance estimate of the difference scores as

$$\hat{\sigma}_\Delta^2 = \frac{\sum_{i=1}^{n}(\Delta_i - \bar{\Delta})^2}{n-1} \tag{3.11}$$

where $\bar{\Delta}$ is the average difference score,

$$\bar{\Delta} = \frac{\sum_{i=1}^{n}\Delta_i}{n}. \tag{3.12}$$

Notice that

$$\bar{\Delta} = \frac{\sum_{i=1}^{n}(Y_i - X_i)}{n} = \frac{\sum_{i=1}^{n}Y_i}{n} - \frac{\sum_{i=1}^{n}X_i}{n} = \bar{Y}-\bar{X}. \tag{3.13}$$

Thus the average difference is equal to the difference of the averages. The t-statistic is then computed as

$$T = \frac{\bar{\Delta}}{\sqrt{\dfrac{\hat{\sigma}_{\Delta}^2}{n}}} = \frac{\bar{Y} - \bar{X}}{\sqrt{\dfrac{\hat{\sigma}_{\Delta}^2}{n}}} \tag{3.14}$$

which is distributed as a Student's t-distribution with n-1 degrees of freedom. Rejection of the null hypothesis indicates there is a significant difference between pretest and posttest scores at level α. If the t-test is tested as a one-sided test, the directionality of the difference can be assessed.

Case 2: The Application of a Treatment Intervention Between Measurement of the Pretest and Posttest Scores

A more common experimental design encountered in clinical research is the case where the experimenter wishes to determine if some treatment is more effective than another. Consider the case where an experimenter measures n subjects twice on a specified variable, once before (pretest) and once after (posttest) an experimental condition called a treatment intervention is imposed on the subjects. This is the exact situation as Case 1 with the exception being this time a treatment intervention is applied between the measurement of the pretest and posttest. In this situation, the linear model for the pretest and posttest may be written as

$$X_i = \mu + S_i + e_{i1} \text{ for the pretest} \tag{3.15}$$

$$Y_i = \mu + S_i + \tau + e_{i2} \text{ for the posttest} \tag{3.16}$$

where all variables are defined the same as earlier and τ is the treatment effect applied between periods. Eqs. (3.15) and (3.16) are identical to Eqs. (3.2) and (3.3), respectively, with the addition of τ in Eq. (3.16). When $\tau = 0$, this is equivalent to the situation above where the question of interest was: "was there a significant difference in pretest and posttest scores?" Now, however, the question of interest is "was there a significant treatment effect, i.e., does τ not equal 0?" The null and alternative hypothesis is:

$$H_o: \tau = 0$$
$$H_a: \tau \neq 0.$$

The goal is to develop a statistic to test whether $\tau = 0$ and see if the null hypothesis is rejected at some nominal α level. It should be pointed out that this type of experimental design is uncontrolled in the sense that τ is not a specific estimator for the treatment effect. τ includes both treatment effect and any uncontrolled variables that were present at the time of the posttest measurement, but not at the time of the pretest. For example, in clinical pharmacology it has recently become known that grapefruit juice dramatically increases the absorption of some drugs, resulting in a larger drug effect due to higher blood drug concentrations. Suppose a study was designed to test the effect of a drug on heart rate. At the time of the pretest, subjects were given a placebo after light breakfast with apple juice (which has no effect on drug absorption), and then had their heart rates were repeatedly measured. At the time of the posttest, all subjects were given the test drug after a light breakfast with grapefruit juice and then had their heart rates repeatedly measured. In this case, τ represents the net effect of drug and grapefruit juice, not solely the effect of the test drug. Thus τ is a biased estimate of the true treatment effect. For this reason, this type of design must be interpreted with caution and it must be assumed that the conditions that were present in Period 1 are exactly the same as in Period 2.

With these caveats in mind, the steps towards developing a test statistic include defining a point estimate for the expected value, defining the variance of the point estimate, and then determining if the point estimate is significantly different from the expected value. Assuming that the true subject effect remains a constant, the difference scores can be expressed as the difference between Eq. (3.16) and Eq. (3.15),

$$Y_i - X_i = \tau + e_{i2} - e_{i1}. \tag{3.17}$$

Assuming the treatment effect is a constant and the expected value of the error terms equals 0, the expected value of the difference scores is

$$E(Y - X) = \tau. \tag{3.18}$$

Thus like Case 1 presented previously, a point estimate for τ is the difference between the posttest and pretest means

$$\hat{\tau} = \overline{\Delta} = \overline{Y} - \overline{X}. \tag{3.19}$$

Because the point estimate is equal to what we wish to estimate (τ), we say that the difference score is an unbiased estimate of the population difference – that is, the estimate is correct "on average" (Olejnik and Algina, 1984). In other words, the average value of the estimate is equal to the average value of the quantity being estimated, assuming that τ is a specific estimator for the true treatment effect and does not include any non-specific variables. Using Eq. (3.9), the standard error may then be written as

$$SE(\tau) = \sqrt{\frac{Var(X) + Var(Y) - 2 \cdot \rho \cdot \sqrt{Var(X)} \cdot \sqrt{Var(Y)}}{n}}. \tag{3.20}$$

Recall that an unbiased estimate of the sample variance for any generic variable, Z, may be computed using the formula

$$Var(Z) = \frac{\sum_{i=1}^{n} (Z_i - \overline{Z})^2}{n - 1}. \tag{3.21}$$

An estimate of the test-retest correlation coefficient may be calculated from the sample

$$r = \frac{\sum_{i=1}^{n}(X_i - \bar{X})(Y_i - \bar{Y})}{\sqrt{\sum_{i=1}^{n}(X_i - \bar{X})^2 \sum_{i=1}^{n}(Y_i - \bar{Y})^2}} \, . \tag{3.22}$$

Eqs. (3.21) and (3.22) may be substituted into Eq. (3.20) to get an unbiased estimate of the standard error of the difference scores. A t-statistic may then be devised using the expected value and variance of the difference scores to test the hypothesis of no difference in pretest-posttest scores

$$t = \frac{\tau}{SE(\tau)} = \frac{\bar{Y} - \bar{X}}{SE(\tau)} \tag{3.23}$$

where $\bar{X}$ and $\bar{Y}$ are the mean pretest and posttest scores, respectively. The t-statistic in Eq. (3.23) has a Student's t-distribution with n-1 degrees of freedom. Rejection of the null hypothesis indicates that a difference exists between pretest and posttest scores.

It can be shown mathematically that Eq. (3.23) is equivalent to Eq. (3.14). To see this numerically, consider the data in Table 3.1. The authors measured the degree of platelet aggregation in 11 individuals before and after cigarette smoking and wished to determine if smoking increases the degree of platelet aggregation. The mean difference was 10.27 with a standard deviation of 7.98. The variance of the before (X) and after (Y) data was 15.61 and 18.30, respectively, and using Eq. (3.22) an estimate of the correlation coefficient between the before and after data was 0.901. Using Eq. (3.14), the t-test statistic is

$$t = \frac{\bar{\Delta}}{\sqrt{\frac{\hat{\sigma}_\Delta^2}{n}}} = \frac{10.27}{\sqrt{\frac{63.62}{11}}} = 4.27 \, .$$

The critical one-sided Student's t-value with 10 degrees of freedom is 1.812. Because the authors phrased the null hypothesis in terms of directionality (one-sided), it may be concluded that smoking increases the degree of platelet aggregation (because the average difference scores was greater than 0) at $\alpha = 0.05$. Using Eq. (3.23), the t statistic, denoted $t(\tau)$ to differentiate it from t, is

$$t(\tau) = \frac{52.45 - 42.18}{\sqrt{\frac{1}{11}\left[18.30^2 + 15.61^2 - 2(0.901)(18.30)(15.61)\right]}} = 4.27 \, ,$$

which is equivalent to Eq. (3.14). Thus the null hypothesis of no difference

$$H_o: \mu_1 - \mu_2 = 0$$

$$H_a: \mu_1 - \mu_2 \neq 0$$

TABLE 3.1

PLATELET AGGREGATION DATA FROM
SMOKERS PRESENTED BY LEVINE (1973)

	Maximum Percent Aggregation		
	Before	After	Difference
	25	27	2
	25	29	4
	27	37	10
	44	f	12
	30	46	16
	67	82	15
	53	57	4
	53	80	27
	52	61	9
	60	59	-1
	28	43	15
Mean	42.18	52.45	10.27
Standard Deviation	15.61	18.30	7.98

is equivalent to the null hypothesis of no treatment effect

$$H_o: \tau = 0$$
$$H_a: \tau \neq 0$$

In summary, two methods have been proposed to analyze data in the one-group case. The first method is the paired samples t-test which collapses the pretest-posttest scores into a single composite measure, the difference score, and then performs a t-test on the difference scores. The other method is to take the difference in the posttest and pretest means and do a t-test taking into account the covariance between pretest and posttest scores. Both methods produce equivalent results.

Nonparametric Alternatives to Case 1 or Case 2

The paired samples t-test assumes that each subject's score is independent of any other subject's scores and that the difference scores are normally distributed. When the assumption of normally distributed test scores is violated, a nonparametric test may be substituted or the p-value can be computed directly from the observed data using randomization techniques (which will be discussed in a later chapter). SAS (1990) provides as part of PROC UNIVARIATE the option NORMAL to test the hypothesis that the data arise from a normal distribution which uses the Shapiro-Wilk W-statistic

(1965) if the sample size is less than or equal to 50 or uses the Kolmogorov D-statistic if the sample size is greater than 50. Another test for normality, the omnibus test for normality, will be presented later in the chapter.

The Wilcoxon (1945) paired-sample test is the nonparametric counterpart to the paired samples t-test and should be used when the assumptions of normality are violated. This test is sometimes called the Wilcoxon signed-rank test or Wilcoxon rank sum test. The only assumption of this test is that the sample distribution is symmetric about the median and that the number of tied ranks is small. The test involves calculating the difference scores, as in the paired samples t-test, and then ranking the difference scores from high to low, affixing the sign of each difference to the corresponding rank. In the case of tied ranks, the mean of the rank which would have been assigned to those observations had they not been tied, is used. For example, if 2 scores each have a value of 8 and the rank of the 2 tied scores is 4 and 5, an average rank of 4.5 is assigned to each score. Differences of 0 are discarded, i.e., they are ignored in the analysis. The ranks which are positive are summed (called T+) and the absolute value of the ranks which are negative (called T-) are summed (hence the term rank sum). The null hypothesis of no difference in pretest and posttest scores is rejected if either T+ or T- is less than or equal to a critical value based on the Wilcoxon test statistic distribution found in most general statistics books. If the one-sided test

$$H_o: \text{pretest scores} <= \text{posttest scores}$$

$$H_a: \text{pretest scores} > \text{posttest scores}$$

is needed, the null hypothesis is rejected if T- is less than or equal the one-tailed Wilcoxon critical value with n degrees of freedom. For the other hypothesis,

$$H_o: \text{pretest scores} >= \text{posttest scores}$$

$$H_a: \text{pretest scores} < \text{posttest scores}$$

the null hypothesis is rejected in T+ is less than or equal to one-tailed Wilcoxon critical value with n degrees of freedom.

Let us look more closely at the data in Table 3.1. The ratio of the variances for the posttest and pretest data was 1.37. Testing the equality of variances of the pretest and posttest data shows that the variances were not equal. However, the difference scores appear highly skewed which may violate the normality assumption of the t-test. Table 3.2 presents the rank sum analysis from the data in Table 3.1. The sum of the positive ranks is 55 and the sum of the negative ranks is 11. Since the sum of the positive ranks and negative ranks is greater than $T_{0.05(2),\ 11}$, we do not reject H_o at $\alpha = 0.05$ and conclude that smoking does not increase the degree of platelet aggregation - a conclusion opposite to that seen using the parametric test. Why were the two conclusions disparate? The parametric test's p-value was invalid because the

TABLE 3.2

WILCOXON SIGNED RANK TEST
FOR DATA IN TABLE 3.1

Before	After	Difference	Rank	Signed Rank
25	27	2	10	10
25	29	4	8.5	8.5
27	37	10	6	6
44	56	12	5	5
30	46	16	2	2
67	82	15	3.5	3.5
53	57	4	8.5	8.5
53	80	27	1	1
52	61	9	7	7
60	59	-1	11	-11
28	43	15	3.5	3.5

$n = 11$
$T+ = 10 + 8.5 + 6 + 5 + 2 + 3.5 + 8.5 + 1 + 7 + 3.5 = 55$
$T- = abs(-11) = 11$
$T_{0.05(2), 11} = 10$
Since T+ is greater than $T_{0.05(2), 11}$ do not reject Ho.

distribution of the difference scores was violated, the assumption of normality. When a statistical tests assumptions are violated the test statistics p-value is not necessarily correct and may be drastically incorrect.

It is commonly accepted and taught in most statistics classes that parametric tests exhibit greater power than their nonparametric counterparts under the normal distribution and when the distribution is not normally distributed, the nonparametric counterpart can attain greater power than the parametric test. Blair and Higgins (1985) published a very interesting paper contradicting this common dictum. The authors used Monte Carlo simulation with finite sample sizes ($n = 10$ per group) to compare the power of the Wilcoxon signed rank test to that of the paired samples t-test under different population shapes. Specifically, the authors compared the power of the tests when both pretest and posttest scores had a distribution that was: normal, uniform, double exponential, truncated normal, exponential, mixed normal, log-normal, chi-square, or Cauchy. As expected, the paired samples t-test had a distinct but small power advantage over the Wilcoxon signed rank test when the data were normally distributed. However, when the data were not normally distributed the Wilcoxon signed rank test was the more powerful test. The authors concluded that in no case did the t-test have more than a small power advantage over the Wilcoxon signed rank test, despite the widely different population distributions studied, and that when the t-test was the more powerful statistic the increase in power was "too small to be of much practical importance." The authors made two summary statements: the claim of parametric tests having greater power than their nonparametric counterpart is often unjustified and that there is no advantage to using the paired samples t-test over the Wilcoxon signed rank test when the distribution of the data is unknown or uncertain. Based on their results, it would appear prudent to include the Wilcoxon signed rank test as part of the analysis and to compare the results of with the paired samples t-test. If the assumptions of the tests are met both the parametric and nonparametric tests should yield approximately equal p-values. If the two results are quite different, then the assumptions of the parametric test must be rigorously examined (if they haven't already). The wisest course of action would be to use the nonparametric test as the primary statistical analysis because of fewer assumptions and almost equal power to the parametric test when the assumptions of the parametric test are met.

Case 3: Two Groups with Different Treatment Interventions Between Measurement of Pretest and Posttest Scores (A Controlled Study Design)

The examples above, and their development, were conditioned on there being only one group with the question of interest being: "was there a significant difference between pretest and posttest scores?" or "did the treatment have a significant effect?" It may be that the researcher wants to compare the results of one treatment with the results of another treatment. In this case the question becomes: "are the treatments equivalent?" The null hypothesis may then be written as:

$$H_o: \quad \tau_1 = \tau_2$$
$$H_a: \quad \tau_1 \neq \tau_2$$

where τ_1 and τ_2 refers to the treatment effect in Group 1 and 2, respectively.

In this type of situation the following design is often used (but not always). Subjects are administered the pretest and then randomized to one of two groups. In one group the treatment of interest is applied, while in the other group another treatment intervention is applied. In one case a neutral intervention is applied which is designed to have no effect, but may not necessarily do so. Alternatively, an active treatment comparison group is used. After the treatment intervention is applied, all subjects are administered the posttest. In this case, the linear model for the pretest scores is given by Eq. (3.15). The linear model for the posttest scores is

$$Y_{1i} = \mu + S_i + \tau_1 + e_{1i}, \quad i=1,2,...n_1 \text{ for group 1} \tag{3.24}$$

$$Y_{2i} = \mu + S_i + \tau_2 + e_{2i}, \quad i=1,2,...n_2 \text{ for group 2} \tag{3.25}$$

where all variables are defined as before. Assuming that S_i has some distribution between individuals, possibly the normal distribution, and is constant over time within an individual, the difference scores can be written as the difference between Eq. (3.24) and Eq. (3.15) for Group 1 and between Eq. (3.25), and Eq. (3.15) for Group 2

$$\Delta_{1i} = Y_{1j} - X_{1i} = \tau_1 + e_{1j} \text{ for Group 1} \tag{3.26}$$

$$\Delta_{2i} = Y_{2i} - X_{1i} = \tau_2 + e_{2i} \text{ for Group 2.} \tag{3.27}$$

Notice that within each group, the difference scores are equivalent to the one-sample case where the

$$E(Y_1 - X_1) = \tau_1 \text{ for Group 1} \tag{3.28}$$

$$E(Y_2 - X_1) = \tau_2 \text{ for Group 2.} \tag{3.29}$$

A natural estimator for τ_1 is $\bar{\Delta}_1$, the average difference for Group 1, while for τ_2, the estimator is $\bar{\Delta}_2$, the average difference for Group 2. If one of the treatments is a control group, $\tau_i = 0$, then the differences in the group mean differences may be used as an estimate of the average treatment effect.

Thus a test for equality of the treatment effects is equivalent to testing for equivalency of the average group difference scores. For two groups, this is equivalent to using the t-test

$$t = \frac{\bar{\Delta}_2 - \bar{\Delta}_1}{\sqrt{\dfrac{s_p^2}{n_1} + \dfrac{s_p^2}{n_2}}} \tag{3.30}$$

where $\bar{\Delta}_1$ and $\bar{\Delta}_2$ are the mean difference scores for Groups 1 and 2, respectively, n_1 and n_2 are the number of subjects in Groups 1 and 2, respectively, and s_p^2 is the pooled variance estimate,

$$s_p^2 = \frac{SS_1 + SS_2}{n_1 + n_2 - 2} = \frac{\sum\limits_{i=1}^{n_1}\left(\Delta_{1i} - \bar{\Delta}_1\right) + \sum\limits_{i=1}^{n_2}\left(\Delta_{2i} - \bar{\Delta}_2\right)}{n_1 + n_2 - 2}, \tag{3.31}$$

where Δ_{1i} and Δ_{2i} are the ith subject's difference score in Groups 1 and 2, respectively. If the observed T is greater than Student's $t_{\alpha(2),n-1}$, the null hypothesis of no treatment effect is rejected.

As an example consider the data in Table 3.3. Subjects participated in a study in which they received a placebo drug capsule and the next day received a sedative capsule. Each subject's psychomotor performance was assessed 8 hr after drug administration on each day using the digit-symbol substitution test (DSST), a subtest of the Weschler Adult Intelligence Scale-Revised. The

TABLE 3.3

**DSST PERFORMANCE BEFORE AND AFTER
ADMINISTRATION OF A SEDATIVE**

Sex	ID	Psychomotor Performance on DSST (# of correct symbol substitutions in 90 sec)		
		No Drug	With Drug	Difference
Males	1	43	40	-3
	2	42	41	-1
	3	26	36	10
	5	39	42	3
	6	60	54	-6
	7	60	58	-2
	8	46	43	-3
	Mean	45.1	44.9	-0.29
	S.D.	12.0	8.0	5.28
Females	11	46	44	-2
	12	56	52	-4
	13	81	81	0
	16	56	59	3
	17	70	69	-1
	18	70	68	-2
	Mean	63.2	62.2	-1.00
	S.D.	12.7	13.2	2.37
t-test		-2.63	-2.90	0.305
Degrees of Freedom		11	11	11
p-value		0.0249	0.0144	0.7664
Rank-transform t-test		-2.496	-3.261	0.2772
p-value		0.0297	0.0076	0.7868

DSST is a pen and paper test whereby a unique set of symbols are associated with digits in the key at the top of the test. The score is the number of symbols correctly drawn in 90 sec. A decrease in score is indicative of psychomotor impairment. Such a long time interval between drug administration and assessment of psychomotor performance may be relevant for drugs that are used as sleeping agents. A patient's psychomotor performance the morning after drug administration may be an issue if the patient will be driving to work or some other task which requires coordination. The researchers were interested in determining whether there was a difference between males and females in the degree of psychomotor impairment.

Since both the pretest (no drug) and posttest (with drug) scores were the same scale, it was considered appropriate to look at the difference in scores. Analysis of the difference scores is useful because it controls for baseline psychomotor performance as well as placebo effect. A negative difference score suggests that the sedative has an negative effect on psychomotor performance. If the researchers ignored the contribution of baseline psychomotor performance to psychomotor performance after drug administration they would conclude that there was a significant difference between males and females ($p = 0.0144$). However there was also a difference in baseline psychomotor performance before any drug was given ($p = 0.0234$) with women scoring better than men on the test. After controlling for baseline differences (by using difference scores) there was no difference between men and women in psychomotor performance after drug administration ($p = 0.7868$). Thus the difference between groups after drug administration was not due to drug administration, but to baseline differences independent of drug administration. Ignoring the influence of baseline differences on psychomotor impairment would have lead to a Type I statistical error, i.e., rejecting the null hypothesis when false.

It is important to stress that the assumptions of the t-test, both in the one- and two-group case, must still be met for the test to be valid. These assumptions are that: each subject's score is independent of the other subject's scores and the distribution of the difference scores must be normally distributed. In the two-group case both groups must also have the same variance. As in the example in Table 3.3, using Shapiro-Wilk W-statistic it was concluded that both the pretest and posttest scores were normally distributed. However, the difference scores were not normally distributed ($p = 0.0349$) at $\alpha = 0.05$, but not at $\alpha = 0.01$. Depending on the level of statistical significance the researcher is comfortable with (and whether the null hypothesis that the data are normally distributed is rejected), a nonparametric alternative may be required.

When the assumptions of the two-group t-test are violated a number of alternative approaches may be used. One method is to compute the difference scores and rank transform the difference scores. A t-test may then be applied to the rank differences. As an example, after converting the data in Table 3.3 to ranks, the resulting t-test was $t(11) = -0.2772$, $p = 0.787$, which was very close to the parametric counterpart $t(11)=0.305$, $p = 0.766$. In this example there was very little difference between the parametric and nonparametric p-value probably because the difference scores were only marginally non-normally distributed. Another alternative is to do a suitable transformation on the difference scores so as to impose normality on the resultant scores.

Case 4: More Than Two Groups with Different Treatment Interventions Between Measurement of Pretest and Posttest Scores (A Controlled Study Design)

Often the researcher may have more than two groups in his experimental design, where one group is a placebo control group and the other groups are different treatments or the same treatment given at different levels. For example, a researcher may want to compare the efficacy of two antipsychotic medications and placebo or two doses of the same antipsychotic medication and placebo. In this instance there are more than two groups and the methods used in Case 1-3 do not apply. When more than two groups are involved, the two-group design can be expanded to a one-way analysis of variance (completely randomized design) on the difference scores because

$$t^2_{\alpha(2),df} = F_{\alpha,1,df} \; .$$

In other words, the square of a t-statistic with df degrees of freedom is equal to the F-distribution with one degree of freedom in the numerator and df degrees of freedom in the denominator.

As an example, consider the data in Table 3.4. Bonate and Jessell (1996) tested freshman college students for their attitudes towards sexual harassment. The higher each student scores, the higher his sensitivity towards sexual harassment. Students were then randomized to one of three treatments: viewing an educational video on sexual harassment, reading literature on sexual harassment, and a neutral control task involving attitudes towards male and female names. A week later students were tested again in their attitudes towards sexual harassment. The posttest scores alone were subjected to analysis of variance and after subtracting the baseline score from each subject, the difference scores were subjected to analysis of variance.

Since both the pretest and posttest scores were measured using the same scale, analysis of difference scores seemed appropriate. A positive difference score after treatment intervention was indicative of an increase in sensitivity towards sexual harassment. Analysis of the posttest scores indicated no difference between groups. However, controlling for the individual differences by including the pretest measurements in the analysis showed that there was a difference among groups (p = 0.0064). Scheffe's test (Kirk, 1982) showed that only the education group was different than the control group and that the educational literature group was different than the video group. The results of the analysis are presented in Table 3.5. There was no evidence of non-normally distributed difference scores, but the residuals showed evidence of non-normality at the 0.05 level (p = 0.0427). Nevertheless, the researchers concluded that educational measures can increase sensitivity towards sexual harassment.

The assumptions of the analysis of variance are slightly different from those of the paired samples t-test. Analysis of variance assumes that each

TABLE 3.4

**SCORES ON THE SEXUAL HARASSMENT INVENTORY
BEFORE AND AFTER TREATMENT INTERVENTION
DATA REPORTED BY BONATE AND JESSELL (1996)**

Summary Statistics	Video Intervention		Education Literature Intervention		Control Group	
	Pretest	Posttest	Pretest	Posttest	Pretest	Posttest
	116	126	185	209	187	199
	183	185	168	189	181	193
	197	162	167	210	147	165
	128	203	193	199	136	120
	182	158	144	177	156	143
	182	161	191	209	160	164
	176	194	174	183	198	212
	182	195	175	175	165	170
	163	170	116	144	185	194
	156	106	146	133	158	151
	202	197	169	184	153	121
	158	186	200	127	167	151
	143	189	176	208	138	148
	152	176	171	135	178	168
	176	201	185	132	140	131
	125	180	187	184	157	154
	167	195	167	200	150	177
	170	199	156	179	105	75
	156	140	185	208	163	177
	171	200	154	188	174	176
	166	169	137	125	165	167
	186	194	145	184	159	164
	187	182	138	172	211	216
	182	166	152	180	171	171
	166	168	125	181	163	144
	188	177	122	169	195	196
	179	204	136	170	190	184
	130	174	140	163	192	188
	163	191	171	180	133	133
	153	131	116	203	189	189
	148	161	134	195		
	181	207	129	183		
	203	184	137	184		
Mean	167	177	157	178	166	165
S.D.	22	24	24	25	23	30
n	33		33		30	

TABLE 3.5

ANALYSIS OF THE SEXUAL HARASSMENT DATA IN TABLE 3.4

ANOVA table for posttest scores					
Source	DF	Sum of Squares	Mean Square	F-Ratio	Prob>F
Treatment	2	3220.05	1610.03	2.30	0.1059
Error	93	65088.91	699.88		
Total	95	68308.96			
ANOVA table for difference scores					
Source	DF	Sum of Squares	Mean Square	F-Ratio	Prob>F
Treatment	2	7052.26	3526.13	5.33	0.0064
Error	93	61493.74	661.22		
Total	95	68546.00			
Means and Standard Errors for Difference Scores					
Group			Mean	Standard Error	
Video Intervention			9.52	4.71	
Education Literature			20.33	5.58	
Control			-0.83	2.45	

posttest score is independent of other posttest scores and identically distributed as the other posttest scores, and the residuals from the analysis of variance are normally distributed with constant variance. The first assumption may be decided by examination of a scatter plot of posttest scores vs. subject number. A significant correlation or pattern in the data is suggestive of dependence between observations. The second assumption may be validated by residual analysis. The reader is referred to a more complete discussion of residual analysis found in any text of regression analysis, such as Neter et al. (1996).

If the assumptions of the analysis of variance are violated, a nonparametric alternative may be substituted, such as the Kruskal-Wallis test (Zar, 1984). Conover and Iman (1981) and Blair and Higgins (1985) have shown that many statistical procedures, such as analysis of variance and analysis of covariance, when performed on ranks, are equal and sometimes outperform their parametric counterparts which are based on raw data.

Unreliability of Difference Scores

The biggest criticism raised in regard to the use of difference scores is that they are unreliable, so much so that analyses based on difference scores have had a difficult time being accepted for publication in some peer-reviewed journals. This criticism is based on research done by Cronbach and Furby (1970) and Lord and Novick (1968), both of whom concluded that difference scores are of little value. The argument that difference scores are unreliable is as follows. The reliability of difference scores, G_d, is given by

$$G_d = \frac{\lambda G_x + \dfrac{G_y}{\lambda} - 2\rho_{xy}}{\lambda + \dfrac{1}{\lambda} - 2\rho_{xy}} \tag{3.32}$$

where G_x and G_y are the individual reliabilities of the pretest and posttest, respectively, ρ_{xy} is the correlation between the observed pretest and posttest scores, and $\lambda = \sigma_x / \sigma_y$ (Guilford, 1954). Note that in this instance, G may not necessarily be equal to ρ, depending on how the reliability of the measuring device was calculated. This formula is based on the assumption that the correlation between pretest and posttest errors is 0. As Williams and Zimmerman (1996) point out, "most textbook authors and others who have contended that gain scores are unreliable have not based their arguments on [Eq. (3.32)], but instead have drawn conclusions from special cases of [Eq. (3.32)]." When $\lambda = 1$, Eq. (3.32) simplifies to

$$G_d = \frac{G_x + G_y - 2\rho_{xy}}{2(1 - \rho_{xy})} \tag{3.33}$$

and when the pretest and posttest are measured using the same measuring device and have equal reliability, G_x, then

$$G_d = \frac{G - \rho_{xy}}{1 - \rho_{xy}}, \quad G > \rho_{xy}, \ \rho_{xy} \neq 0. \tag{3.34}$$

Obviously when the correlation between scores is equal to their individual reliabilities, $G = \rho_{xy}$, the reliability of the difference scores is equal to 0 and completely unreliable. What might not seem as obvious is that for any G and ρ_{xy}, G_d is always less than G (Figure 3.1). Thus difference scores are said to be unreliable. This argument has stood for almost a quarter of a century before being challenged.

Recently, however, researchers have begun to question this dogma for a number of reasons. Foremost, if the analysis of data using difference scores results in a significant finding, does the unreliability of difference scores imply the finding is invalid? Of course not. Williams and Zimmerman (1996) have made a quite persuasive argument which says that although difference scores are unreliable, the difference in reliability between the individual pretest and posttest reliabilities can be small. They state that the dogma of unreliability of

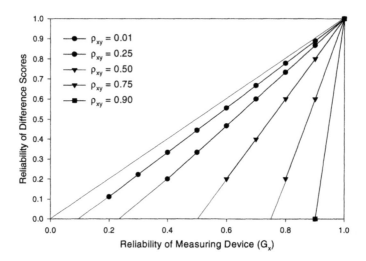

Figure 3.1: Reliability of difference scores as function of the reliability of the measuring device and correlation between pretest and posttest scores using Eq. (3.34). The dashed line is the line of unity. For any combination of G_x or ρ_{xy}, the reliability of difference scores is less than the individual reliabilities.

difference scores is always based on the worst case scenario; that those authors who argue that difference scores are unreliable use assumptions that "show the reliability of differences in the most unfavorable light" (Williams and Zimmerman, 1996). Under the assumptions made in developing Eq. (3.34), the reliability of difference scores will be at a minimum. It is not that Williams and Zimmerman (1996) are arguing that difference scores are reliable, they are arguing that it is wrong to simply conclude that difference scores will be unreliable under any and all conditions. They also argue that it is too restrictive to assume that the standard deviation of the pretest and posttest scores are always equal, especially after intervention of a treatment intervention. The reliability of difference scores must be viewed as a composite function of all its components, not as a simplification. When viewed as a whole, the usefulness of difference scores becomes more apparent.

Distribution of Difference Scores

One advantage to using difference scores is that if the marginal distributions of the pretest and posttest are normally distributed, the distribution of the difference scores will be as well. As will be seen in the next chapter, this is not the case with relative change scores. A univariate test for normality having good power at detecting departure from normality is the omnibus test for normality (D'Agostino, Belanger, and D'Agostino, 1990). If X is an n x 1 vector of scores, such as difference scores, calculate the skewness, $\sqrt{b_1}$, and kurtosis, β_2 , as

$$\sqrt{b_1} = \frac{m_3}{\sqrt{m_2^3}} \tag{3.35}$$

$$\beta_2 = \frac{m_4}{m_2^2} \tag{3.36}$$

where

$$m_k = \frac{\sum_{i=1}^{n}(X_i - \overline{X})^k}{n}$$

and n is the sample size. Compute

$$Y = \sqrt{b_1}\sqrt{\frac{(n+1)(n+3)}{6(n-2)}} \tag{3.37}$$

$$\beta_2\left(\sqrt{\beta_1}\right) = \frac{3\left(n^2 + 27n - 70\right)(n+1)(n+3)}{(n-2)(n+5)(n+7)(n+9)} \tag{3.38}$$

$$W^2 = -1 + \sqrt{2\beta_2\left(\sqrt{\beta_1}\right) - 1} \tag{3.39}$$

$$\delta = \frac{1}{\sqrt{Ln(W)}} \tag{3.40}$$

$$\theta = \sqrt{\frac{2}{W^2 - 1}} \tag{3.41}$$

$$Z_1 = \delta \cdot Ln\left[\frac{Y}{\theta} + \sqrt{\left(\frac{Y}{\theta}\right)^2 + 1}\right]. \tag{3.42}$$

Under the null hypothesis that the sample comes from a normal distribution, Z_1 is approximately normally distributed. Eq. (3.37) to (3.42) are based on the skewness of the samples. The other half of the omnibus test is based on the kurtosis. Compute

$$E(\beta_2) = \frac{3(n-1)}{n+1} \qquad (3.43)$$

$$Var(\beta_2) = \frac{24n(n-2)(n-3)}{(n+1)^2(n+3)(n+5)} \qquad (3.44)$$

$$q = \frac{\beta_2 - E(\beta_2)}{\sqrt{Var(\beta_2)}} \qquad (3.45)$$

$$\sqrt{\beta_1(\beta_2)} = \frac{6(n^2-5n+2)}{(n+7)(n+9)}\sqrt{\frac{6(n+3)(n+5)}{n(n-2)(n-3)}} \qquad (3.46)$$

$$A = 6 + \frac{8}{\sqrt{\beta_1(\beta_2)}}\left[\frac{2}{\sqrt{\beta_1(\beta_2)}} + \sqrt{1+\frac{4}{\beta_1(\beta_2)}}\right] \qquad (3.47)$$

$$Z_2 = \frac{\left[\left(1-\frac{2}{9A}\right) - \sqrt[3]{\frac{1-\frac{2}{A}}{1+q\sqrt{\frac{2}{A-4}}}}\right]}{\sqrt{\frac{2}{9A}}}. \qquad (3.48)$$

Under the null hypothesis that the sample comes from a normal distribution, Z_2 is approximately normally distributed. The omnibus test statistic is then calculated as

$$K^2 = Z_1^2 + Z_2^2. \qquad (3.49)$$

Under the null hypothesis, K^2 has a chi-squared distribution with 2 degrees of freedom.

Effect of Regression Towards the Mean on Difference Scores

Much was said about regression towards the mean and difference scores in the last chapter, but we will now examine it in a little more greater detail. Difference scores transform pretest-posttest data into a univariate data set. Ideally, the transformed variable is independent of the component scores (Kaiser, 1989). The analysis may then proceed using the resultant transformed variable. Difference scores are sometimes criticized because they are not independent of their component scores and are often correlated with the pretest score. As stated many times earlier, subjects with pretest scores greater than the mean will tend to have smaller difference scores and, conversely, subjects with pretest scores below the mean will tend to have larger difference scores. In this sense, difference scores are biased. The correlation between pretest and difference scores, $\rho_{x,y-x}$, is given by

$$\rho_{x,y-x} = \frac{\sigma_x \sigma_y G - \sigma_x^2}{\sigma_x \sigma_{y-x}} \tag{3.50}$$

where σ_{y-x}, σ_x and σ_y are the standard deviations of the difference, pretest and posttest scores, respectively, and σ_x^2 is the variance of the pretest scores (Ghiselli, Campbell, and Zeddeck, 1981). When measurements are made using the same instrument and $\sigma_x = \sigma_y$, then Eq. (3.50) may be simplified to

$$\rho_{x,y-x} = \frac{\sigma_x^2 (G-1)}{\sigma_x \sigma_{y-x}}. \tag{3.51}$$

Only when the denominator of Eq. (3.51) is large compared to the numerator and the test-retest correlation between pretest and posttest scores is very close to one does the correlation between pretest scores and difference scores become close to 0. Because the numerator in Eq. (3.50) is the difference between the pretest/posttest covariance and the variance of the pretest scores, it is common for $\rho_{x,y-x}$ to be negative because the variance of the pretest scores is often greater than the pretest/posttest covariance. As we will see in later chapters (Analysis of Covariance), a researcher may take advantage of the correlation between pretest and difference scores by using the pretest score as a covariate in an analysis of covariance while using the difference score as the dependent variable.

The problem with the correlation between difference and pretest scores is that regression towards the mean significantly influences the value of the corresponding difference score. Explicit in the development of the paired samples t-test is that regression towards the mean is not occurring. For the t-test to be valid when significant regression towards the mean is occurring, adjusted posttest scores, as given in Eq. (2.32), should be used instead of raw posttest scores. In the case where subjects are enrolled in a study on the basis of their pretest scores, the influence of regression towards the mean becomes magnified. If μ is known then an alternative solution to the t-test problem is presented by Mee and Chua (1991). They presented a regression-based test to be used when regression towards the mean is occurring to a significant extent. Let $Z_i = X_i - \mu$. Assuming $Z_i = z_i$ is fixed then

$$Y = \mu + \tau + \rho z + e \tag{3.52}$$

where e is random error with mean 0 and variance σ^2. As can be seen, Eq. (3.52) indicates that Y and z are related in a linear manner with the slope equal to the test-retest correlation coefficient between Δ and z and intercept equal to the sum of the population mean and treatment effect. Hence, the null hypothesis of treatment effect is equivalent to testing the null hypothesis for the intercept in the linear regression model. Let β_1 and β_0 be the ordinary least squares (OLS) estimates for the slope and intercept, respectively, of Δ regressed against z. Assuming that the standard deviation of the pretest equals the standard deviation of the posttest, a test of H_0 is given by

$$t = \frac{\beta_0 - \mu}{\left[\left\{ MSE \left[\frac{1}{n} + \frac{\bar{z}^2}{\sum_{i=1}^{n}(z_i - \bar{z})^2} \right] \right\} \right]^{1/2}}$$

(3.53)

where MSE is the means square error obtained from the linear regression of Δ on z, μ is the population mean, and the t-value follows a Student's t-distribution with n-2 degrees of freedom.

Use of this method will be demonstrated using the example they presented. A group of students must pass an examination in order to receive high school diplomas. If any students fail they may take a refresher course and retake the exam. Data for eight students are shown in Table 3.6. Using a paired t-test, the t-value is 2.0 with a p-value of 0.0428. Thus one might conclude that there was a differential effect of the refresher course on the exam. However, as mentioned above, the paired t-test assumes that regression towards the mean is not occurring. Given that the population mean is 75, $\bar{X} = -17.625$, $\hat{\beta}_1 = 1.11$, $\hat{\beta}_0 = 79.96$, MSE $= 20.30$, and $\sum(X_i - \bar{X})^2 = 341.88$, the t-value for testing H_o: $\beta_0 = 75$ vs. H_a: $\beta_0 > 75$ is

$$t = \frac{79.96 - 75}{\left\{ 20.296 \left[0.125 + \frac{17.625^2}{341.875} \right] \right\}^{1/2}} = \frac{4.96}{4.58} = 1.08$$

which has a p-value of 0.16. Thus after correcting for regression towards the mean, it was concluded that the refresher course had no effect. Their method has some disadvantages, which Mee and Chua (1991) point out. First (and this applies to most statistical tests), if the linear model is incorrectly specified or the data are not normally distributed then this test statistic is invalid. Second, this test is sensitive to differential treatment effects. For example, those subjects who failed dismally may show better improvement than subjects who were only slightly below the cutoff for failure. Third, μ must be known with certainty. Mee and Chua (1991) present some suggestions for how to correct for these problems. The reader is referred to their paper for further details.

TABLE 3.6

DEMONSTRATION OF REGRESSION BASED t-TEST
FOR TESTING H_0: $\tau = 0$
DATA PRESENTED BY MEE AND CHUA (1991)

Student	Pretest (X)	Posttest (Y)	$Z = X-\mu$	Difference
1	45	49	-30	4
2	52	50	-23	-2
3	63	70	-12	7
4	68	71	-7	3
5	57	53	-18	4
6	55	61	-20	6
7	60	62	-15	2
8	59	67	-16	8
Mean	57.4	60.4	-17.6	4
S.D.	7.0	8.8	7.0	3.2

RESULTS OF REGRESSION ANALYSIS FROM SAS

Dependent Variable: POSTTEST

Analysis of Variance

```
                  Sum of      Mean
Source    DF     Squares     Square      F Value      Prob>F
Model      1    422.09877   422.09877    20.797       0.0038
Error      6    121.77623   20.29604
Total      7    543.87500

Root MSE          4.50511    R-square       0.7761
Dep Mean         60.37500    Adj R-sq       0.7388
C.V.              7.46188
```

Parameter Estimates

```
Parameter                  Standard   T for H0:
Variable   DF  Parameter   Error      Parameter=0   Prob > |T|
INTERCEPT  1   79.95905    4.580258   17.457        0.0001
Z          1   1.111152    0.243653   4.560         0.0038
```

George et al. (1997) present a more complicated maximum likelihood approach for when μ is not known. Using Monte Carlo simulation they showed that their method has greater power than Mee and Chua's (1991) method under certain conditions. The reader is referred to that paper for further details.

Summary

- Difference scores:
    1. Attempt to remove the influence of the pretest score on the posttest score but often fail in this regard because difference scores are usually negatively correlated with pretest scores.
    2. Require that the pretest and posttest be collected with the same measuring device and have the same units.
    3. Are easy to interpret and easily analyzed. This is the main reason they are used.
    4. Are less than (usually) or equal to (rarely) in reliability as the individual component scores.
- Nonparametric counterparts to the parametric statistical methods presented have only slightly less power when the assumptions of the parametric test are met and greater power when the assumptions of the parametric test are violated.
    ◊ It is suggested that most analyses be done using nonparametric methods unless one is confident in the assumptions of the parametric test.
- When the marginal distributions of the pretest and posttest are normally distributed, the distribution of their difference scores will be normally distributed.
- Beware of the impact of regression towards the mean in the analysis of difference scores and use analysis of covariance or corrected difference scores when necessary.

CHAPTER 4

RELATIVE CHANGE FUNCTIONS

As mentioned in the last chapter, relative change scores are another commonly used method to control for the influence of the pretest score on the posttest score. Since there are a variety of different types of relative change functions, the definition of each will be presented individually.

Definitions and Assumptions

Relative change scores require the pretest and posttest scores to be continuous random variables, thus ensuring the change score to be a continuous random variable. Relative change scores using interval data make little sense since the resulting change score then becomes an interval. Relative change scores also require the pretest and posttest variables to be the same type of measurement made using the same device and have equal units of measurement. Although pretest and posttest scores have the same units, relative change scores are often unitless or expressed as percentages. It is not necessary for either the pretest, posttest, or relative change score to be normally distributed for their use. In fact, the biggest drawback of relative change scores is that they are often not normally distributed.

Relative change scores convert the pretest and posttest scores into a proportional change score, C, expressed as either raw change or absolute change. The formula to convert pretest and posttest scores can be written as

$$C = \frac{Y - X}{X} \qquad (4.1)$$

in the case of raw change, or

$$|C| = \frac{|Y - X|}{X} \qquad (4.2)$$

in the case of absolute change, where C is the change score, Y is the posttest score, and X is the pretest score. Note that the numerator is a difference score, whereas the denominator scales the difference score. A variant of these equations is to multiply the proportional change scores by 100 thereby converting them to percent change scores. If $C = 0$, no change has occurred. A positive relative change score indicates that the posttest score was greater than the pretest score, whereas a negative relative change score indicates that the posttest score was less than the pretest score. One criticism of relative changes scores is in the choice of the scaling term or denominator. Consider an individual whose initial score is 3 and whose final score is 7. Using Eq. (4.1), this represents a 133% increase from baseline. However, if a patient scores a 7 initially and deteriorates to a 3, a -57% decrease has occurred.

Hence, different denominator terms result in different transformations and estimates of change.

Proportional and percent changes scores fall under a family of transformations known as change functions. Tornqvist, Vartia, and Vartia (1985) formally defined a change function as a real-value function $C(Y, X)$ of positive arguments, $C: R_+^2 \to R$, with the following properties:

1. $C(Y,X) = 0$,   if $Y = X$.
2. $C(X,Y) > 0$,   if $Y > X$.
3. $C(X,Y) < 0$,   if $Y < X$.
4. C is a continuous increasing function of Y when X is fixed.
5. $\forall a{:}a > 0 \to C(aY,aX) = C(Y,X)$.

The last property merely states that the function is independent of units of measurement. The property $C: R_+^2 \to R$ states that a two-dimensional vector $\left(R_+^2\right)$ is mapped into a one-dimensional vector (R) by the function C. It can be shown that both proportional and percent change functions meet these requirements. It can also be shown that difference scores represent another valid type of change function.

As with difference scores, it is useful to examine the expected value and variance of the proportional change scores. The expected value of Eq. (4.1) can be written as

$$E(C) = E\left(\frac{Y-X}{X}\right) = E\left(\frac{Y}{X} - \frac{X}{X}\right) = E\left(\frac{Y}{X}\right) - E\left(\frac{X}{X}\right)$$

$$E(C) = E\left(\frac{Y}{X}\right) - 1. \tag{4.3}$$

Thus the expected value of a change score simplifies to finding the expected value of the ratio of Y to X. The expected value of Y/X can be written as a first-order Taylor series expansion

$$E\left(\frac{Y}{X}\right) = \frac{\mu_y}{\mu_x} + \frac{1}{\mu_x^2}\left(\sigma_x^2 \frac{\mu_y}{\mu_x} - \rho\sigma_x\sigma_y\right). \tag{4.4}$$

Because $E\left(\dfrac{Y}{X}\right)$ does not equal μ_y/μ_x, Eq. (4.1), is biased, i.e., the average value of the estimate is not equal to the average value of the quantity being estimated. The second term in Eq. (4.4) is the degree of bias in the proportional change score which depends on several factors, including the mean and standard deviation of the pretest and posttest and the reliability of measuring device.

Using the rules of expectation, the variance of Eq. (4.1) can be written as

$$\text{Var}\left(\frac{Y-X}{X}\right) = \text{Var}\left(\frac{Y}{X} - \frac{X}{X}\right) = \text{Var}\left(\frac{Y}{X} - 1\right) = \text{Var}\left(\frac{Y}{X}\right) - 1$$

$$\text{Var}\left(\frac{Y-X}{X}\right) = \frac{1}{\mu_x^2}\left(\sigma_x^2\frac{\mu_y^2}{\mu_x^2} + \sigma_y^2 - 2\rho\sigma_x\sigma_y\frac{\mu_y}{\mu_x}\right) \tag{4.5}$$

Thus the variance of the proportional change score is dependent on the same factors that affect the expectation of the change score. It should be stressed that the same results hold for when change is expressed as a percentage. In this case, Eq. (4.4) and Eq. (4.5) are expressed as

$$E(C\%) = 100\left[\frac{\mu_y}{\mu_x} + \frac{1}{\mu_x^2}\left(\sigma_y^2\frac{\mu_y}{\mu_x} - \rho\sigma_x\sigma_y\right) - 1\right] \tag{4.6}$$

and

$$\text{Var}\left(\frac{Y-X}{X} \times 100\%\right) = \frac{100^2}{\mu_x^2}\left(\sigma_x^2\frac{\mu_y^2}{\mu_x^2} + \sigma_y^2 - 2\rho\sigma_x\sigma_y\frac{\mu_y}{\mu_x}\right) \tag{4.7}$$

respectively.

Tornqvist, Vartia, and Vartia (1985) have shown that most every indicator of relative change can be expressed as a function of Y/X alone. Hence, the change function can be expressed as an alternate function dependent solely on Y/X. Formally, there exists a function H, such that

$$C(X,Y) = H\left(\frac{Y}{X}\right) = C\left(\frac{Y}{X}, 1\right) \tag{4.8}$$

with properties:

1. $H\left(\frac{Y}{X}\right) = 0$, if $\frac{Y}{X} = 1$.

2. $H\left(\frac{Y}{X}\right) > 0$, if $\frac{Y}{X} > 1$.

3. $H\left(\frac{Y}{X}\right) < 0$, if $\frac{Y}{X} < 1$.

4. H is a continuous increasing function of its argument Y/X.

5. $H\left(\frac{aY}{aX}\right) = aH\left(\frac{Y}{X}\right)$.

Given that the expected value of Y/X is biased, any relative change function that can be expressed as a function H will have some degree of bias due to the expected value of Y/X. Table 4.1 shows a variety of other relative

TABLE 4.1

RELATIVE CHANGE FUNCTIONS AND THEIR SIMPLIFICATION INTO FUNCTIONS OF X_2/X_1 AS PRESENTED BY TORNQVIST, VARTIA, AND VARTIA (1985)

Function	Mapping
$\dfrac{X_2 - X_1}{X_1}$	$\dfrac{X_2}{X_1} - 1$
$\dfrac{X_2 - X_1}{X_2}$	$1 - \dfrac{X_1}{X_2}$
$\dfrac{X_2 - X_1}{(X_2 + X_1)\big/2}$	$\dfrac{\left(\dfrac{X_2}{X_1} - 1\right)}{\dfrac{1}{2}\left(1 + \dfrac{X_2}{X_1}\right)}$
$\dfrac{X_2 - X_1}{\sqrt{X_2 X_1}}$	$\dfrac{\dfrac{X_2}{X_1} - 1}{\sqrt{\dfrac{X_2}{X_1}}}$
$\dfrac{X_2 - X_1}{\left[\dfrac{1}{2}\left(X_2^{-1} + X_2^{-1}\right)^{-1}\right]}$	$\dfrac{\left(\dfrac{X_2}{X_1} - 1\right)\left(1 + \dfrac{X_1}{X_2}\right)}{2}$
$\dfrac{X_2 - X_1}{\min(X_2, X_1)}$	$\dfrac{\dfrac{X_2}{X_1} - 1}{\min\left(1, \dfrac{X_2}{X_1}\right)}$
$\dfrac{X_2 - X_1}{\max(X_2, X_1)}$	$\dfrac{\dfrac{X_2}{X_1} - 1}{\max\left(1, \dfrac{X_2}{X_1}\right)}$
$\dfrac{X_2 - X_1}{K(X_1, X_2)}$ where K is any mean of X_2 or X_1	$\dfrac{\dfrac{X_2}{X_1} - 1}{K\left(1, \dfrac{X_2}{X_1}\right)}$

change functions proposed by Tornqvist, Vartia, and Vartia (1985) and their simplification into functions of Y/X.

The ideal relative difference function has two additional properties

$$H\left(\frac{Y}{X}\right) = -H\left(\frac{X}{Y}\right) \qquad (4.9)$$

$$H\left(\frac{Z}{X}\right) = H\left(\frac{Y}{X}\right) + H\left(\frac{Z}{Y}\right). \qquad (4.10)$$

Eq. (4.9) states that the resulting relative change distribution is symmetric about its mean, while Eq. (4.10) states that the relative change function is additive. Neither proportional change scores nor percent changes have these properties.

Statistical Analyses with Change Scores

Both difference scores and change scores map a two-dimensional random variable into a one-dimensional random variable. Thus change scores can be used with any of the statistical tests developed for difference scores with the caveat that the assumptions of the test are verified for the test to be valid.

Change Scores and Regression Towards the Mean

As one might expect, since change scores are scaled difference scores they are not immune from the influence of regression towards the mean. Indeed, change scores are often highly correlated with baseline values and in this regard the difficulties that are present in the analysis of difference scores are the same for the analysis of change scores. The data shown in Figure 2.1 are plotted as percent change scores in Figure 4.1. The negative correlation between difference scores and pretest scores is also seen between percent change scores and difference scores. Not surprisingly, the correlation between difference and pretest scores is equal to the correlation between percent change and difference scores. In both cases, $r = -0.35$ ($p < 0.0001$).

It is sometimes thought that the problems inherent in difference score analysis are avoided using percent change scores, but this is a fallacy. Analysis of change scores has all the problems encountered with difference scores plus others that are unique to change scores alone. Assuming that the linear model for posttest scores can be written as

$$Y_i = \mu + S_i + \tau + e_i \qquad (4.11)$$

where τ is the treatment effect (a constant which affects all individuals equally), then like difference scores, relative change scores can be corrected for the influence of regression towards the mean using the following equation

$$C_{adj} = \frac{Y - X}{X} - \frac{\sigma_y (\rho - 1)}{\sigma_x}\left(1 - \frac{\mu}{x}\right). \qquad (4.12)$$

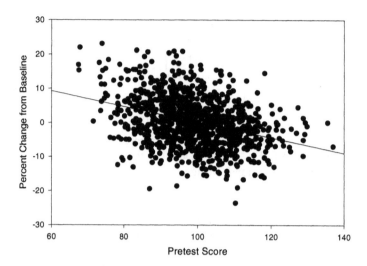

Figure 4.1: Plot of percent change from baseline against baseline scores from data in Figure 2.1. The solid line is the least-square linear regression line to the data. The negative correlation is evidence that regression towards the mean occurs even with percent change scores.

Similarly adjusted percent change scores can be calculated by multiplying Eq. (4.12) by 100% (Chuang-Stein, 1993). Figure 4.2 shows Figure 4.1 after adjusting for regression towards the mean. The observed correlation coefficient between pretest scores and posttest scores is 0.0327, which is non-significant, indicating the corrected change scores are independent of pretest scores.

Difference Scores or Relative Change Scores?

Choosing whether to use difference scores or percent change scores in an analysis depends on many factors. One factor often used is dependent on which variable will be easier for the reader to understand. It would be difficult for a non-clinician to interpret a change of 20 in his cholesterol level, but telling him that his cholesterol dropped 10% is easier to understand. Another factor is which variable is impacted to a greater extent by regression towards the mean. Kaiser (1989) suggests using the variable which has less correlation with pretest scores. Kaiser (1989) also proposed a more formal test statistic upon which to make a decision and showed that the ratio of the likelihoods comparing difference scores to percent change scores can be used as a criterion measure for selecting which test statistic to use. For k groups and j subjects in each group, the ratio of the maximum likelihoods is

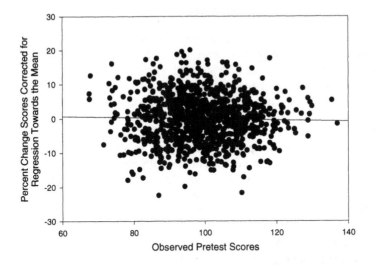

Figure 4.2: Plot of adjusted percent change scores against their baseline score using Eq. 4.12. Compare to figure 4.1. There is no correlation between baseline scores and adjusted posttest scores indicating that regression towards the mean has been corrected for.

$$R = \left[\frac{\left(geo(\overline{X})\right)^2 \sum\limits_{ij} \left[\frac{\left(C\% - \overline{C}\%_i\right)}{100} \right]^2}{\sum\left(D_{ij} - \overline{D}_i\right)^2} \right]^{n/2} \tag{4.13}$$

I = 1, 2, ... k, j = 1, 2, ...n, where $geo(\overline{X})$ is the geometric mean of the pretest scores,

$$geo\left(\overline{X}\right) = \sqrt[n]{X_1 X_2 ... X_n} = \sqrt[n]{\prod_{i=1}^{n} X_i} \tag{4.14}$$

$\overline{C}\%_i$ is the ith group's average percent change score, and $\overline{D}_i$ is the ith group's average difference score. If R is greater than one, use difference scores, otherwise use percent change scores. Although this test statistic was developed for normal populations, using computer simulation, Kaiser (1989) indicated that it works well with populations exhibiting substantial positive skewness. Kaiser points out, however, that R not be solely relied on in choosing to use percent change or difference scores, but to use the judgment of the researcher and analyst.

Going back to the platelet aggregation data by Levine (1973), the geometric mean of the pretest score was 39.5, the mean percent change was 26.5, and the ratio, R, was 0.75 suggesting that percent change scores should be used in the statistical analysis. However, this is in contrast to examination of the scatter plots which suggests that difference scores should be used in the analysis. Closer examination of the correlation coefficients shows that the standard error of the correlation coefficient for *both* plots of difference score and percent change vs. pretest scores were small enough that significance testing of the correlation coefficient for both plots failed to reject the null hypothesis; the correlation coefficient for both plots was not significantly different from 0. This example demonstrates that neither technique should be used alone, but should be used in conjunction with the other.

Other Relative Change Functions

A variety of other functions have been proposed to assess change. All of these methods share a data transformation approach in common, wherein pretest and posttest scores are collapsed into a single summary measure. Though all of these will not be reviewed in detail, two additional ones will be. Berry (1990) and Brouwers and Mohr (1989) proposed using a modification of the proportional change equation, Eq. (4.1), which we will call the modified proportional change score[*]:

$$C_{mc} = \frac{Y - X}{\left[\dfrac{(Y + X)}{2}\right]} \tag{4.15}$$

where the denominator of Eq. (4.1) is replaced by the average of the pretest and posttest scores. It can be seen that Eq. (4.15) is equivalent to one of the functions in Table 4.1 originally presented by Tornqvist, Vartia, and Vartia (1985). Brouwers and Mohr (1989) argue that the advantage of using modified proportional change over the traditional proportional change score is that the transformed variable does not depend on the denominator used in the transformation and the resultant distribution is symmetrical about its mean. Figure 4.3 plots the data in Figure 2.1 using the modified proportional change function. The bottom of Figure 4.3 shows that indeed the data are symmetrical about the mean and normally distributed. However, modified percent change

---

[*] Berry (1990) refers to this as a symmetrized change score.

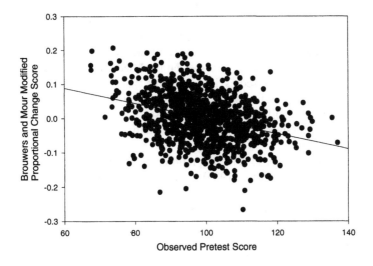

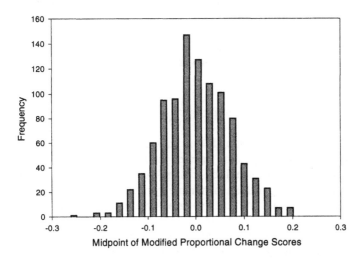

Figure 4.3: Modified proportional change scores as suggested by Brouwers and Mohr (1989) plotted against the baseline score (top) using the data in Figure 2.1. The solid line is the least-square linear regression line to the data. Even though this transformation tends to result in modified change scores that are normally distributed (bottom), the negative correlation between modified change score and baseline score indicates that modified proportional change scores are also subject to regression towards the mean.

scores are still correlated with pretest scores and are therefore also subject to regression towards the mean (top of Figure 4.3). Also, as will be seen in the next subsection, although modified change scores improve the sampling distribution properties by making them more symmetrical compared to percent change scores, there are still many cases where they are not normally distributed. A big disadvantage in using the modified change score is in its interpretation. A subject's scores increase from 50 at pretest to 150 upon posttest, 3 times their baseline score. This is readily understood by anyone. However, using the modified metric this represents a 50% increase which is vastly different from what someone would intuitively think of as a 50% increase.

Another proposed metric is the log ratio or log difference score proposed by Tornqvist, Vartia, and Vartia (1985). Log ratio scores are computed as

$$L = Ln\left(\frac{Y}{X}\right) = Ln(Y) - Ln(X), \quad X, Y \geq 0. \tag{4.16}$$

Log ratio scores have the advantage of being symmetric and are normalized to pretest scores. Also, defining the metric in this manner avoids the issue of which denominator term should be used. Log ratio scores have the advantage in that they are the only metric which is additive in nature. Log ratio scores are the only change function which has both the ideal properties of a change function: symmetry and additivity. The authors argue that Eq. (4.16) can be multiplied by 100 to convert the scores to log percent scores. An example of log ratio scores is found in acoustics where the change in decibels is expressed as

$$dB = 10 \times Log_{10}\left(\frac{P_1}{P_0}\right) \tag{4.17}$$

where P_1 is some measurement and P_0 is background. Log difference scores are, however, subject to regression towards the mean. As will be seen in the next subsection, although modified change scores improve the sampling distribution properties by making them more symmetrical compared to percent change scores, there are still many cases where they are not normally distributed.

It is apparent that log-ratio scores are valid only if the pretest and posttest scores are positive, nonzero numbers because the natural log of a negative number or zero does not exist. However, one method which can be used if any pretest or posttest measurements are negative or 0 is to add a constant to each score, such that the log-ratio score becomes

$$L = Ln\left(\frac{Y}{X}\right) = Ln(Y + c) - Ln(X + c) \tag{4.18}$$

where c is a constant which forces X and Y to be greater than 0. Berry (1987) presents a method that may be used to define the optimal value of c while still ensuring a symmetrical distribution. The skewness, $\sqrt{\beta_1}$, and kurtosis, β_2 , of a distribution were introduced in the last chapter as a way to test the distributional properties of difference scores. Herein we will use these statistics to identify a transformation constant, c, such that the new data set will have better normality properties than the original scores. Let $\sqrt{\beta_1}$ and β_2 be defined by Eq. (3.35) and (3.36), respectively. If we define

$$ g = \left| \sqrt{\beta_1} \right| + \left| \beta_2 \right| \qquad (4.19) $$

then the optimal value of c is the value of c which minimizes g. The value of c can then be found using either a numerical optimization routine or a one-dimensional grid search.

As an example, consider the artificial data presented in Table 4.2. In this hypothetical example, subjects are administered either a drug or placebo and asked the question are they tired or awake on a visual analog scale. The scale ranges from "tired" with a score of -50 to "wide awake" with a score of 50. Thus the interval presented to each subject is {-50, 50} with a range of 100 units. In this example, both positive, negative, and 0 scores are present. The minimum value that c can be is greater than 14 because if c is less than 14, some scores will still be less than or equal to 0. Therefore, c is constrained to be greater than 14. Figure 4.4 plots g as a function of c using a grid search technique. It can be seen that the optimal value of c was 22.8. This resulted in a skewness of -0.152 and a kurtosis of -3.00. The bottom plot in Figure 4.4 is the resultant distribution of the modified log ratio scores. The log-ratio scores can now be used as statistics in either a t-test or analysis of variance. The SAS code used to obtain g is given in the Appendix.

Distribution of Relative Change Scores

In the last chapter it was noted that when the distributions of the pretest and posttest were normally distributed, the distribution of the difference scores was normally distributed as well (see Chapter 3: Distribution of Difference Scores). Does this mean that their corresponding relative change scores will also be normally distributed? A simulation was conducted to test the hypothesis that if the pretest and posttest are normally distributed, then the distribution of the relative change scores will also be normally distributed. Pretest-posttest data were generated having a specific test-retest correlation with a specific sample size (see Chapter 8: Generation of Bivariate, Normally Distributed Data with a Specified Covariance Structure) were generated. The sample sizes ranged from 10 to 200 and the correlation ranged from 0 to 0.95. The percent change scores, log-ratio scores using Eq. (4.16), and modified change scores using Eq. (4.15) were calculated for each data set and were tested for normality using the omnibus test for normality as presented by D'Agostino, Belanger, and D'Agostino (1990). The critical value used was 0.05 and 1000 iterations were used for each sample-size correlation

TABLE 4.2

**HYPOTHETICAL EXAMPLE OF SUBJECTIVE
EFFECTS OF DRUG AND PLACEBO**

Subject	Placebo	Drug	Subject	Placebo	Drug
1	-10	25	20	0	34
2	5	28	21	2	32
3	7	45	22	3	16
4	-8	32	23	-6	17
5	0	4	24	-7	-1
6	-14	18	25	10	11
7	2	15	26	8	14
8	-10	14	27	0	25
9	12	-4	28	1	19
10	7	9	29	-4	18
11	18	24	30	2	16
12	-11	23	31	-4	15
13	-1	10	32	2	7
14	0	16	33	5	-1
15	-2	18	34	-1	3
16	4	2	35	0	-2
17	7	8	36	13	27
18	10	16	37	9	26
19	-4	24	38	5	32

combination. The average percent of simulations declaring the distribution of the relative change scores to be normally distributed as shown in Figures 4.5-4.7.

For percent change scores, a test-retest correlation and sample size interaction was apparent. For any given sample size, there was a distinct trend that as the test-retest correlation increased, the percent of simulations declaring the percent change scores to be normally distributed increased. For any given test-retest correlation, as the sample size increased, the percent of simulations declaring the percent change scores to be normally distributed decreased dramatically. The overall result indicating that if the distribution of the pretest and posttest scores were normally distributed, the distribution of the percent change scores will tend not to be normally distributed with increasing sample size and decreasing test-retest correlation.

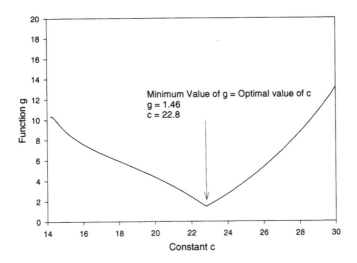

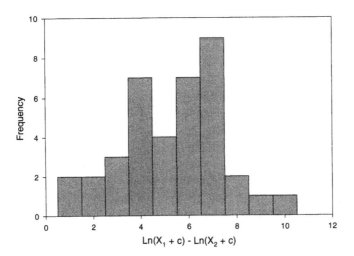

Figure 4.4: Here, g (Eq. 4.19) is seen as a function of constant c using the log-ratio score with adjustment as proposed by Berry (1987). The minimum value of g is the optimal value of the constant to add to each score. The bottom plot shows the distribution of the log-ratio scores after log-transformation. After adding a value of 22.8 to each value and using the log-ratio score as the primary metric, the data are normally distributed (bottom).

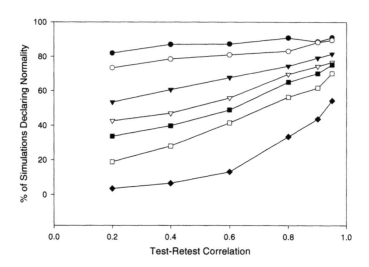

Figure 4.5: Percent of simulations declaring the distribution of percent change scores to be normally distributed. Legend: solid circle, n = 10; open circle, n = 20; solid upside-down triangle, n = 40; open upside-down triangle, n = 60; solid square, n = 75; open square, n = 100; solid diamond, n = 200.

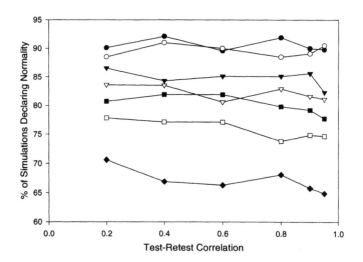

Figure 4.6: Percent of simulations declaring the distribution of log-ratio scores to be normally distributed. Legend: solid circle, n = 10; open circle, n = 20; solid upside-down triangle, n = 40; open upside-down triangle, n = 60; solid square, n = 75; open square, n = 100; solid diamond, n = 200.

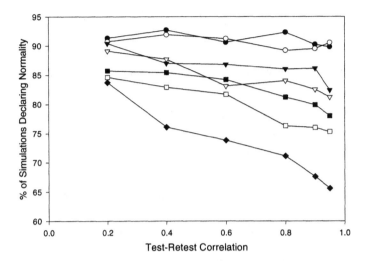

Figure 4.7: Percent of simulations declaring the distribution of modified percent change scores to be normally distributed. Legend: solid circle, n = 10; open circle, n = 20; solid upside-down triangle, n = 40; open upside-down triangle, n = 60; solid square, n = 75; open square, n = 100; solid diamond, n = 200.

For both log-ratio and modified change scores, no interaction between test-retest correlation and sample size was observed. Instead, as the test-retest correlation increased, the percent of simulations declaring the data to be normally distributed remained relatively constant. Within any given test-retest correlation, as the sample size increased, the percent of simulations declaring the data to be normally distributed decreased, although not nearly as dramatically as with percent change scores.

For any combination of sample size and test-retest correlation, the percent of simulations declaring log-ratio and modified change scores to be normally distributed was greater than percent change scores, indicating that these transformations have better distributional properties than percent change scores. These results show that with increasing sample size, the distribution of relative change scores tended to be non-normally distributed, even if the marginal scores were normally distributed. With percent change scores, the situation is even more complicated.

Summary

- Relative change scores are commonly found in the literature, both as descriptive statistics and in use in hypothesis testing.

- Relative change functions convert pretest and posttest scores into a single summary measure which can be used with any of the statistical tests developed for difference scores.
- All the problems seen with difference scores are seen with relative change scores. In addition
    1. The average percent or average proportional change score is a biased estimator of the population change.
    2. The degree of bias is dependent on many variables including the correlation between pretest and posttest scores and the initial pretest score. It should be pointed out, however, that if the average pretest score is large, the degree of bias may be small.
- Converting pretest and posttest scores into a relative change score may result in a highly skewed or asymmetrical distribution which violates the assumptions of most parametric statistical tests.
- Change scores are still subject to the influence of regression towards the mean.
- It is the opinion of the author that percent change scores should be avoided. If their use cannot be avoided, check to make sure the scores do not violate the assumptions of the statistical test being used.
    - If the relative change score is normally distributed than an appropriate transformation should be used to convert the data to normality.
- Use of log-ratio scores appear to be a viable alternative to traditional change scores because they are not as asymmetrical as percent change scores.
- The distribution of percent change scores tends to be not normally distributed with increasing departure from normality as the sample size increases and with decreasing departure from normality as the test-retest correlation increases. Both log-ratio scores and modified percent change scores have better symmetry properties than percent change scores and they tend to be normally distributed more often than percent change scores.

CHAPTER 5

ANALYSIS OF COVARIANCE

Recall in Chapter 1 that many researchers feel that hypothesis testing for baseline comparability is redundant because the very act of hypothesis testing assures that some percentage of studies will be declared baseline non-comparable. It is generally recognized, however, that if baseline non-comparability among groups is suspected then analysis of covariance (ANCOVA) is the recommended method of analysis (Senn, 1994; Overall, 1993). Also, recall from Chapter 2 that when regression towards the mean significantly influences the measurement of the posttest, then ANCOVA is the recommended method of choice. For these and other reasons, many researchers advocate ANCOVA as the method of choice for pretest-posttest data analysis.

Definitions and Assumptions

Analysis of covariance (ANCOVA) combines regression with analysis of variance and can be thought of as a statistical control technique because ANCOVA in pretest-posttest analyses treats the pretest score as a covariate, a continuous variable that represents a source of variation which has not been controlled for in the experiment and is believed to affect the expected value of the posttest score (Kirk, 1982). One advantage to using ANCOVA is that unlike difference scores or relative change scores, the pretest need not be measured using the same device as the posttest scores. The only requirement is that the pretest and posttest scores are correlated in some manner.

ANCOVA adjusts the dependent variable so as to remove the influence of the pretest on the posttest and basically answers the question: are the treatment effects different between groups when applied to individuals *with the same baseline pretest score*? The interpretation of ANCOVA vs. analysis of variance is subtle. Whereas analysis of variance tests the hypothesis that the mean posttest scores among treatment groups are equal, ANCOVA tests the hypothesis that the mean posttest score in each group is equal given that the mean pretest score in each treatment group is equal. Many texts and reviews have been written on ANCOVA and it is beyond the scope of this chapter to give an extensive treatise on the subject. If the reader is interested, he may refer to Wildt and Ahtola (1978) or Kirk (1982). Nevertheless, the basic assumptions of ANCOVA and the special issues in using ANCOVA in the analysis of pretest-posttest designs will be presented.

Parametric ANCOVA

For k groups with j subjects in each group, the simple linear model for ANCOVA is

$$Y_{ij} = \mu + \tau_i + \beta_w \left(X_{ij} - \overline{X}_{..} \right) + e_{ij}, \quad j=1,2..k \tag{5.1}$$

where Y_{ij} is the jth subject in the ith's groups posttest score, μ is the population grand mean, τ_i is the ith group's treatment effect, X_{ij} is the jth subject in the ith group's pretest score, β_w is the common linear regression coefficient, e_{ij} is random error, and $\overline{X}_{..}$ is the grand mean pretest score. Crager (1987) has shown that in the case of pretest-posttest designs, the between-groups regression coefficient is a function of the between subject, σ_S^2, and within subject error variances, σ_e^2,

$$\beta_w = \frac{\sigma_S^2}{\sigma_S^2 + \sigma_e^2} \tag{5.2}$$

such that β_w is always constrained to the interval (0,1). It is readily apparent that β_w is the test-retest reliability coefficient and it can now be seen why ANCOVA is the method of choice when regression towards the mean significantly influences the posttest scores. ANCOVA explicitly models the influence of the pretest scores on the posttest scores by inclusion of a term for regression towards the mean in the linear model.

Compared to a completely randomized design where

$$Y_{ij} = \mu + \tau_i + e_{ij} \tag{5.3}$$

it can be seen that ANCOVA splits the error term into two components: one due to regression of the pretest scores on the posttest scores and another due to unexplained or residual variation. Hence, the mean square error term is smaller in ANCOVA than in a completely randomized design and provides a better estimate of the unexplained variation in the model when β_w is significantly different than 0. An alternative way of viewing ANCOVA may be that ANCOVA proceeds as a two-step process. In the first step, ANCOVA adjusts the posttest scores by removing the influence of the pretest scores, such that

$$Y_{(adj)ij} = Y_{ij} - \beta_w \left(X_{ij} - \overline{X}_{..} \right) = \mu + \tau_i + e_{ij} \tag{5.4}$$

where $Y_{adj(ij)}$ is the adjusted posttest score. The second step is to perform an analysis of variance on the adjusted posttest scores. It can be shown that when β_w is at its upper limit, i.e., $\beta_w = 1$, both ANCOVA and ANOVA using difference scores will have similar sum of squares values in the denominator of the F-ratio, but ANOVA on difference scores will be slightly more powerful due to a single lost degree of freedom associated with the regression component in the ANCOVA. However, when $\beta_w \neq 1$ (as is usually the case when the pretest and posttest are in fact correlated), the error term in the

ANCOVA will be smaller in ANOVA using difference scores. Thus under most conditions the ANCOVA model will provide greater statistical power than analysis of variance using difference scores (Huck and McLean, 1975).

Cochran (1957) showed that the reduction in the mean square error term using ANCOVA compared to a completely randomized experimental design is determined by the size of the test-retest correlation between the pretest and posttest scores, ρ. If $\sigma^2_{e,anova}$ is the error term when covariance adjustment is not used as in a analysis of variance, then the error term when covariance, $\sigma^2_{e,ancova}$, is used is given by

$$\sigma^2_{e,ancova} = \sigma^2_{e,anova} \left(1 - \rho^2\right)\left(1 + \frac{1}{df_e - 2}\right) \tag{5.5}$$

where df_e is the degrees of freedom associated with $\sigma^2_{e,anova}$. Normally the degrees of freedom is sufficiently large that the reduction in the error term depends primarily on the correlation coefficient between pretest and posttest scores with a higher correlation resulting in a smaller mean square error.

ANCOVA with Difference Scores as the Dependent Variable

Recall from Chapter 3 that when the pretest and posttest are measured using the same measuring instrument and reported with the same units, the difference in the posttest and pretest scores are often negatively correlated with pretest scores. It has been suggested that an alternative analysis is to treat the pretest scores as covariates but to use difference scores, as opposed to posttest scores, as the dependent variable. Either method is valid depending on whether the assumptions of the analysis of variance are valid with regard to the data.

However, Laird (1983) has shown that estimation of the treatment effect is independent of using either raw scores or difference scores as the dependent variable – both methods give exactly the same result. To see this, if n individuals are treated in k groups, for the simple ANCOVA using the pretest score as the covariate, the estimated treatment effect, $\hat{\tau}_i$, is calculated as

$$\hat{\tau}_i = \overline{X}_{2i.} - \overline{X}_{2..} - \beta_w\left(\overline{X}_{1i.} - \overline{X}_{1..}\right) \tag{5.6}$$

where $\overline{X}_{2i.}$ is the mean for the ith groups posttest score, i = 1, 2, ...k, $\overline{X}_{2..}$ is the grand mean posttest score, $\overline{X}_{1i.}$ is the mean for the ith groups pretest score, and $\overline{X}_{1..}$ is the grand mean pretest score. If an ANCOVA is performed on the difference scores, the estimated treatment effect, $\tilde{\tau}_i$ is given by

$$\tilde{\tau}_i = \overline{d}_{i.} - \overline{d}_{..} - \tilde{\beta}_w\left(\overline{X}_{1i.} - \overline{X}_{1..}\right). \tag{5.7}$$

Solving for $\left(\overline{X}_{1i.} - \overline{X}_{1..}\right)$ and substituting into Eq. (5.6), it can be seen that $\tilde{\beta}_w = \beta_w - 1$ and $\hat{\tau}_i = \tilde{\tau}_i$. As an example, consider the data in Table 3.4. An ANCOVA on the difference and posttest scores is presented in Table 5.1. Notice that the F-ratios for treatment effect and pretest by treatment interaction are exactly the same in both analyses. Similarly, the mean square error is also the same in both analyses. The only difference is in the estimation of the pretest effect. This example illustrates that ANCOVA on either posttest or difference scores yields identical analyses.

ANCOVA Using Percent Change as the Dependent Variable

Some researchers have used percent change as the dependent variable in an ANCOVA with the pretest scores as the covariate, an analysis that will be called percent change ANCOVA. Suissa, Levinton, and Esdaile (1989) have shown that using percent change ANCOVA results in a quadratic relationship, not a linear relationship, between baseline and outcome. To see this, consider the case where there is no treatment intervention applied between measurements of the pretest and posttest, the linear model (written in linear regression notation) would be

$$\frac{Y_i - X_i}{X_i} \times 100\% = \beta_0 + \beta_1 \cdot X_i + e_i \tag{5.8}$$

where Y_i and X_i are the posttest and pretest scores, respectively, for the ith subject, β_0 is the intercept (which is an estimate of μ), β_1 is the slope of the relationship between pretest and posttest, and e is random error. If we redefine $\beta_j' = \beta_j / 100$, $j = 0$ and 1, then the expected value of the posttest scores can be written as

$$E(Y) = \left(1 + \beta_0'\right) \cdot X + \beta_1' \cdot X^2. \tag{5.9}$$

Thus the expected posttest score is a quadratic function of the pretest score. This function will either be concave up or down depending on whether $\beta_1 < 0$ or $\beta_1 > 0$. When $\beta_1 < 0$ the extrema, X^*, can be found at

$$X^* = \frac{1 + \beta_0}{2 \cdot \beta_1}. \tag{5.10}$$

Figure 5.1 demonstrates the relationship in Eq. (5.9) for both positive and negative β_1. Because the expected value function is dependent on β_1, two possible interpretations arise. One interpretation is that when $\beta_1 > 0$, there is a positive relationship between pretest and posttest for all pretest scores. The other is that when pretest scores are less than the extrema and $\beta_1 < 0$ then there

TABLE 5.1

**ANALYSIS OF THE SEXUAL HARASSMENT
DATA PRESENTED IN TABLE 3.4 USING ANCOVA**

ANCOVA Table for Posttest Scores with Pretest Scores as a Covariate

Source	DF	Sum of Squares	Mean Square	F-Ratio	Prob>F
Treatment	2	11689.3	5944.6	12.82	0.0001
Pretest	1	15574.3	15574.3	34.17	0.0001
Interaction	2	9907.5	4953.7	10.87	0.0001
Error	90	41025.3	455.8		
Total	95	78196.4			

ANCOVA Table for Difference Scores with Pretest Scores as a Covariate

Source	DF	Sum of Squares	Mean Square	F-Ratio	Prob>F
Treatment	2	11689.3	5944.6	12.82	0.0001
Pretest	1	9206.1	9206.1	20.20	0.0001
Interaction	2	9907.5	4953.7	10.87	0.0001
Error	90	41025.3	455.8		
Total	95	71828.2			

is a positive relationship between pretest and posttest. However, when the pretest scores are greater than the extrema and $\beta_1 < 0$, there is a negative relationship between pretest and posttest. Thus it is possible for some subjects to have a positive relationship between pretest and posttest whereas others may have a negative relationship. Because two possible relationships may arise with percent change ANCOVA, it is possible for mixed results to occur. An example using percent change ANCOVA and the possible misinterpretation that may result is the report by Whiting-O'Keefe et al. (1982) who showed that a high inverse serum creatinine level was a negative prognostic factor for lupus nephritis. This is in contrast to other reports which found that low inverse creating levels were negative prognostic factors for lupus nephritis (Fries et al., 1978; Austin et al., 1983). Suissa, Levinton, and Esdaile (1989) suggest that the authors' conclusion was due to the potential misinterpretation of percent change ANCOVA. Because different interpretations can arise in percent change ANCOVA simply due to the sign of one of the regression coefficients, their use should be discouraged.

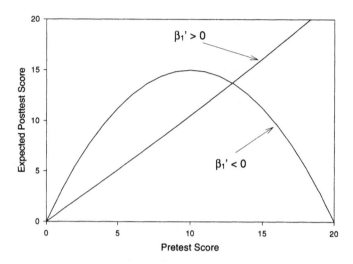

Figure 5.1: Expected posttest score as a function of baseline score under an ANCOVA model using percent change as the dependent variable and baseline score as the covariate. There is a quadratic relationship between expected value and baseline score when $\beta_1' < 0$, where β_1' is the regression parameter for the quadratic term in the linear model relating expectation to baseline score (see Eq. 5.9). When $\beta_1' > 0$, the relationship is linear. Reprinted from *The Journal of Clinical Epidemiology*, 42, Suissa, S., Levinton, C., and Esdaile, J.M., Modeling percentage change: a potential linear mirage, 843-848, Copyright (1989), with permission from Elsevier Science.

Assumptions of the ANCOVA

In order for the omnibus F-test for a significant treatment effect in ANCOVA to be valid certain assumptions must be met. These are

1. The errors are independent and normally distributed with mean 0 and common constant variance.
2. The population within-groups regression coefficients are equal, i.e., $\beta_1 = \beta_2 = ...\beta_k = \beta_w$.
3. The pretest scores are measured without error.
4. If a curvilinear relationship exists between the pretest and posttest scores, that relationship is reflected in the linear model.

In general, ANCOVA is relative robust to deviations from assumption of normality and homogeneity of variance (Assumption 1), as is the usual ANOVA (Atiqullah, 1964). However, the remaining assumptions are often violated in common practice with pretest-posttest data. How to deal with the effect of each of these violations on the analysis will now be presented.

Violation of Homogeneity of Within-Groups Regression Coefficients

The assumption that the within-groups regression coefficients are equal is often violated. A quick test for heterogeneity of regression coefficients is to perform using an ANCOVA with the inclusion of a term for the group by pretest interaction. Interaction reflects an unequal treatment effect between groups, and if the interaction is significant, the assumption of homogeneity of regression coefficients is violated. In general, ANCOVA with unequal slopes is analogous to a two-way analysis of variance with significant interaction between the factors (Feldt, 1958). Figure 5.2 plots how the group by pretest interaction appears in a scatter plot of pretest vs. posttest scores. Clearly the slopes of the two groups are unequal. Thus as pretest scores increase along the ordinate, the difference between pretest and posttest scores becomes larger. This is the differential nature of interaction. Jaccard (1990; 1997) presents a more comprehensive discussion for interaction in ANOVA and regression models. The reader is referred therein for more in-depth detail.

When the group by pretest interaction term is statistically significant, there are several methods for dealing with heterogeneity of the regression coefficients. These include Quade's nonparametric ANCOVA (1967), Puri and Sen's nonparametric ANCOVA (1969), and parametric ANCOVA applied to the ranks (Conover and Iman, 1982; Olejnik and Algina, 1984; Seaman, Algina, and Olejnik, 1985). These tests are designed to ignore the assumption of homogeneity of regression coefficients and yet still provide sensitive tests for significant treatment effects.

Quade's (1967) procedure is to first transform the pretest and posttest scores to ranks by ranking each pretest and posttest score. Note that separate ranks are computed for both pretest and posttest scores. The ranks of both the pretest and posttest scores are then converted to deviation scores from the mean rank, d_{pre} and d_{post}, respectively, by subtracting the mean rank from each observation. Ordinary least squares linear regression is then used to estimate the regression coefficient, $\hat{\beta}$, which predicts d_{post} as a function of d_{pre}. This is equivalent to computing the Spearman rank correlation coefficient on the raw ranks. The deviation rank of the posttest, $\hat{d}_{post}$, is then predicted from the deviation rank of the pretest by

$$\hat{d}_{post} = d_{pre} \times \hat{\beta}. \tag{5.11}$$

The predicted posttest deviation rank is then subtracted from the observed deviation rank of the posttest to form a residual deviation rank score, e_r,

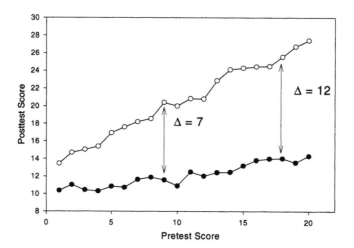

Figure 5.2: Plot showing how group by pretest interaction appears when posttest scores are plotted against pretest scores. No interaction would appear as parallel lines and a constant difference between groups for a given pretest score. Interaction appears as diverging lines and non-constant differences for a given pretest score. Shown in the plot is that the difference between the two groups is 7 units when the pretest score was about 10, but the difference is 12 when the pretest score was about 18.

$$e_r = \hat{d}_{post} - \text{Rank(Posttest)}. \tag{5.12}$$

An analysis of variance is then performed on the residual deviation rank scores, e_r. Basically, Quade's procedure is to compute a linear regression on the mean corrected ranks, then subject the residuals to an analysis of variance. Puri and Sen (1969) developed a test statistic, L_n, which tests the hypothesis of no treatment effect. L_n is distributed as a chi-square random variable and Quade's method is a special case of their more general method, although the approximating distributions are different.

Lastly, parametric ANCOVA on the ranked pretest and posttest scores without regard to treatment group has been used because rank transformation methods often have properties of robustness and power in both regression methods and analysis of variance. Therefore, it seems reasonable that rank transformation in ANCOVA will also be robust and have a high degree of statistical power. Using Monte Carlo simulation, Conover and Iman (1982) showed that the Type I error rate remains approximately α for both the parametric rank method and Quade's method when the posttest scores have either a log-normal, exponential, uniform, or Cauchy distribution, thus indicating robustness of ANCOVA against non-normality; the parametric rank method loses power when then posttest scores are distributed as log-normal or Cauchy distributions; and the analysis of ranks appears to be better when the posttest scores are distributed as skewed exponential or log-normal distributions, whereas analysis of the raw data is better when the posttest scores are normally or uniformly distributed.

Olejnik and Algina (1984) and Seaman, Algina, and Olejnik (1985) did a similar study to Conover and Iman wherein they studied the power and Type I error rate of rank-transformed ANCOVA to parametric ANCOVA using pretest scores as the covariate. In this study the authors varied the sample size between groups, the correlation between pretest and posttest, the conditional distribution of the residuals and covariate, and combinations thereof. In general, Type I error rates for both tests were near their nominal values. Parametric ANCOVA was shown to be robust to deviations in both normality of residuals and homoscedasticity among treatment groups. Rank-transform ANCOVA was also robust to deviations from normality, but when the sample sizes were small, the data were both non-normal and heteroscedastic, and there was a weak correlation between pretest and posttest, rank-transform ANCOVA was liberal in declaring statistical significance. However, under all other conditions, rank-transform ANCOVA was robust to deviations from assumptions. Seaman, Algina, and Olejnik (1985) conclude that "rank ANCOVA has Type I error rates near the nominal level, is usually more powerful than the parametric ANCOVA, and can be substantially more powerful (when the assumptions of parametric ANCOVA are violated)." These studies clearly demonstrate the superiority of rank-transform ANCOVA over parametric ANCOVA, especially when certain assumptions for parametric ANCOVA are violated.

As an example of these methods, Conover and Iman (1982) present a modified data set originally presented by Quade (1967). The Appendix presents the SAS code to compute a test for heterogeneity of variance, Quade's test, standard ANCOVA, and ANCOVA on the ranks. Table 5.2 presents the data and Table 5.3 presents the results from the various tests. The test for heterogeneity of regression coefficients was nonsignificant ($p = 0.333$), so the hypothesis of equal regression coefficients was not rejected. Both nonparametric methods gave similar p-values which were slightly different from the parametric ANCOVA method, although neither of the nonparametric methods rejected the hypothesis of a difference between groups. Notice that Quade's procedure had different degrees of freedom than standard ANCOVA, but Quade's gave an F-value almost exactly that of parametric ANCOVA on the ranks. This is in agreement with the simulation study by Conover and Iman (1982) who demonstrated that Quade's test and parametric ANCOVA on the ranks are usually in virtual agreement.

TABLE 5.2

**DATA PRESENTED BY CONOVER AND IMAN (1982) TO
DEMONSTRATE VARIOUS NONPARAMETRIC ANCOVA
PROCEDURES**

Group	Posttest	Pretest	Rank (Posttest)	Rank (Pretest)
1	16	26	1	7
	60	10	5	3
	82	42	7	11
	126	49	11	13
	137	55	12	14
2	44	21	4	6
	67	28	6	8
	87	5	8	2
	100	12	9	4
	142	58	13	15
3	17	1	2	1
	28	19	3	5
	105	41	10	10
	149	48	14	12
	160	35	15	9

Courtesy of Conover, W.J. and Iman, R.L., Analysis of covariance using the rank transformation, *Biometrics* 38, 715-724, 1982. With permission from the International Biometrics Society.

TABLE 5.3

RESULTS OF ANCOVA ANALYSIS OF TABLE 5.2

Test	F-value	Degrees of freedom	p-value
Heterogeneity of variance (Treatment by group interaction)	1.25	2, 9	0.333
Test for treatment effect using parametric ANCOVA without a treatment by group interaction term	0.74	2,11	0.498
Quade's test for treatment effect	1.30	2,12	0.307
Test for treatment effect using parametric ANCOVA on ranks without a treatment by group interaction term	1.30	2,11	0.312

Hamilton (1976) used Monte Carlo simulation to examine the power of parametric ANCOVA and the three nonparametric methods presented above when the regression slopes among the groups were unequal. All the methods were quite robust to deviations from homogeneity of variance when there were equal sample sizes in each group and the posttest scores were normally distributed. Neither parametric ANCOVA nor the nonparametric methods were robust when the sample sizes were unequal among groups. Also, the chi-square approximation proposed by Puri and Sen for their test statistic, L_n, was poor for small sample sizes, $n < 10$ per group. As expected, when the assumptions for ANCOVA were met and the sample size was relatively large, $n > 10$ per group, parametric ANCOVA had larger statistical power than the nonparametric methods.

Error-in-variables ANCOVA

Another assumption often violated in ANCOVA is the assumption of error-free measurement of the pretest scores. As mentioned in the second chapter, no measurement is error free; all measurements have some degree of random or systemic error. Often in the case of pretest-posttest data, the degree of random error is the same for both the pretest and the posttest. It might not be apparent that the error term in the linear model

$$Y_{ij} = \mu + \tau_i + \beta_w \left(X_{ij} - \overline{X}_{..} \right) + e_{ij}, \ j=1,2..k \qquad (5.13)$$

represents the random component of Y_{ij}; it does not reflect measurement error in X_{ij}. In the case where significant error is made in the measurement of the pretest the linear model must be modified by inclusion of an additional error term, e_{xij}, to reflect this error

$$Y_{ij} = \mu + \tau_i + \beta_w \left(X_{ij} - \overline{X}_{..} + e_{xij} \right) + e_{ij}, \ j=1,2..k . \qquad (5.14)$$

By expanding terms, it can be shown that

$$\text{Cov}(X, \ e - \beta_w d) \neq 0 ,$$

which violates the assumption that random error term is independent of other terms in the linear model. The net effect of errors in the measurement of the pretest is that it attenuates the estimation of β_w, obfuscates true treatment effects, and creates the illusion of differences which may or may not be present. Lord (1955, 1963) showed that as the error in the measurement of the pretest increases, the increased precision in the estimation of the mean square error seen with ANCOVA decreases and ANCOVA becomes more and more like the corresponding ANOVA.

This problem is often called error-in-variables regression. Two simple nonparametric solutions to this problem were presented by Knoke (1991). An alternative solution to the error-in-variables problem, one that is much more complex than the solution presented by Knoke (1991), was presented by DeGracie and Fuller (1972). The reader is referred to the paper for further details. Referring back to the paper by Knoke (1991), he proposed that one method to control for error-in-variables is to rank transform the pretest and posttest data and submit the ranks to an ANCOVA. The other method suggested by Knoke (1991) is to compute the normalized rank scores from the ranks and submit these scores to an ANCOVA. Normalized rank scores are approximations to the expected order statistics assuming the normal distribution. Two such normalized rank transformations are based on Blom (1958) and Tukey (1962)

$$R_{z(i)} = \frac{\Psi\left[R_i - \frac{3}{8}\right]}{n + \frac{1}{4}} \text{ for Blom} \qquad (5.15)$$

$$R_{z(i)} = \frac{\Psi\left[R_i - \frac{1}{3}\right]}{n + \frac{1}{3}} \text{ for Tukey} \qquad (5.16)$$

where $R_{z(i)}$ is the rank normal score, R_i is the rank of X, n is the number of non-missing observations of X, and $\Psi[.]$ is the inverse cumulative normal distribution function. Note that SAS can provide these values as part of the PROC RANK procedure using the NORMAL= option. Focus will primarily center on Blom's transformation because it has been shown to fit better to the normal distribution than Tukey's transformation. As an example, consider the data set {27, 31, 45, 67, 81} with ranks {1, 2, 3, 4, 5}. The normalized rank scores using Blom's transformation is {-1.18, -0.50, 0, 0.50, 1.18}. For any vector of scores, normalized rank scores center the ranks at 0 and force them to be symmetrical around 0.

Knoke (1991) used Monte Carlo simulation to investigate the performance of the usual parametric ANCOVA based on the raw data and the nonparametric alternatives under the error-in-variables model. He examined the influence of various measurement error distributions: normal, uniform, double exponential, log-normal, Cauchy, etc. with varying degrees of skewness and kurtosis on power and Type I error rate. The results indicated that an increase in the error variance in the measurement of the pretest decreased the power of both the parametric and nonparametric tests. When the error variance was not large and the data were normally distributed, there was little loss of power using the normal rank scores transformation compared to parametric ANCOVA. Rank transformed scores tended to have less power than both normal rank transformed scores and parametric ANCOVA. When the measurement error was non-normally distributed, the normal rank scores had much greater efficiency than the other methods, although as the sample size increased or as the separation between groups increased, the efficiency of using rank transformed scores approached the efficiency of using normal rank transformed scores. Knoke (1991) concluded that normal rank scores transformation be used in practice when there is considerable error in the pretest measurement.

Other Violations

If a curvilinear relationship exists between pretest and posttest scores, the general ANCOVA linear model may be modified to include higher order polynomial terms

$$Y_{ij} = \mu + \tau_i + \beta_1\left(X_{ij} - \overline{X}_{..}\right) + \beta_2\left(X_{ij} - \overline{X}_{..}\right)^2 + e_{ij} \tag{5.17}$$

where β_1 and β_2 are the first order and second order common regression coefficients. If the relationship between pretest and posttest scores is nonlinear, then perhaps nonlinear mixed effects models may be needed. The reader is referred to an excellent text by Davidian and Giltinan (1995) on practical use of these nonlinear models.

Effect of Outliers and Influential Observations

An often undiscussed subject in ANCOVA is when outliers or influential observations are present in the data. Outliers, which will be referred to as influential observations, are those "observations that are not consistent with the proposed model in that these observations differ from the 'bulk' of the data and may arise through measurement or recording error, typographical error, or a specification error in the linear model" (Birch and Myers, 1982). The impact of influential observations may be that they bias the estimate of the regression parameter, β_w, and inflate the estimate of the mean square error.

It is beyond the scope of this book to discuss methods for outlier detection suffice to say that standard methods which are used in linear regression to detect influential observations can also be used in ANCOVA. Thus PRESS residuals, studentized residuals, Cook's distance, DFFITS, and DFBETAs may be used to determine the degree of leverage or influence an observation may exert on the model and overall goodness of fit. SAS provides each of these diagnostics in PROC GLM which may be accessed using the OUTPUT statement. The reader is referred to Neter et al. (1996) for further reading on these methods.

Under normal conditions, the solution to any linear model is by ordinary least squares (OLS) which minimizes the sum of the squared differences between the observed and model predicted values

$$\min \sum_{i=1}^{q} \sum_{j=1}^{n} \left(Y_{ij} - \hat{Y}_{ij}\right)^2 \tag{5.18}$$

where

$$\hat{Y}_{ij} = \hat{\mu} + \hat{\tau}_i + \hat{\beta}_w\left(X_{ij} - \overline{X}_{..}\right) + e_{ij} \tag{5.19}$$

$\hat{\mu}$ is the estimated population grand mean, $\hat{\tau}_i$ is the ith group's estimated treatment effect, and $\hat{\beta}_w$ is the estimated regression coefficient between the pretest and posttest scores.

It is clear from Eq. (5.18) that an outlier of the type where a particular Y_{ij} is different from the rest of the values in the vector Y will have significant influence on the least-squares estimates and the overall mean square error or residual variance. Hence, least-squares estimates are said to be non-robust to outliers in that inaccurate and imprecise parameter estimates can result. One method used to minimize the influence of an observation on parameter estimation is iteratively reweighted least-squares (IRWLS). A summary of the steps are as follows:

1. Choose a weight function to weight all observations.
2. Obtain starting weights for all observations.
3. Use the starting weights and perform weighted least squares.
4. Obtain the residuals.
5. Use the residuals to obtain revised weights.
6. Go to step 3 and continue until convergence is achieved.

Many weight functions have been proposed to deal with outliers. The most commonly used weight functions are the Huber function or bisquare function. These are:

Huber:
$$w = \begin{cases} 1 & |u| \le 1.345 \\ \dfrac{1.345}{|u|} & |u| > 1.345 \end{cases} \qquad (5.20)$$

bisquare:
$$w = \begin{cases} \left[1 - \left(\dfrac{u}{4.685} \right)^2 \right]^2 & |u| \le 4.685 \\ 0 & |u| > 4.685 \end{cases} \qquad (5.21)$$

where w is the weight and u denotes the scaled residual

$$u_i = \dfrac{e_i}{\left[\dfrac{1}{0.6745} \, \text{median}\{ |e_i| \} \right]} \qquad (5.22)$$

and e_i is the ith residual. The constant 1.345 in the Huber function and 4.685 in the bisquare function are called tuning constants. The denominator in Eq. (5.22) is referred to as the median absolute deviation estimator. Seber (1977) recommends that two iterations be used in computing the parameter estimates under IRWLS due to convergence difficulties on successive iterations.

Using Monte Carlo simulation, Birch and Myers (1982) compared the efficiency of ordinary least squares (OLS) to the bisquare and Huber functions in ANCOVA. The authors concluded that the bisquare function was superior to the Huber function when the data had a large proportion of outliers (> 10%), otherwise the estimators were nearly equal. The efficiency of the two methods increased as the sample size increased, where the term efficiency refers to the ratio of the variance of parameter estimates using IRWLS to the variance of parameter estimates using OLS. The authors concluded that both OLS and IRWLS "be used together to provide a basis of comparison and diagnostic examination of the data."

As an example of how OLS estimates differ from IRWLS estimates consider the data in Table 5.4. In Group 2 one subject has a posttest score of 19, possibly as a result of a typographical error, which is clearly discrepant to the other observations. Table 5.5 presents the results from the ANCOVA using iteratively reweighted least-squares and ordinary least squares with and without the outlier. The Appendix presents the SAS code to perform the analysis. Analyzing the data using iteratively reweighted least-squares resulted in the detection of a significant treatment effect compared to OLS estimates. The overall treatment F-value produced by the Huber function was very close to the F-value produced by OLS with the outlier deleted from the analysis, but was different than the F-value produced by the bisquare function. In no instance did the pretest have predictive power over the posttest score. Also, notice the difference in degrees of freedom using the bisquare function. This difference resulted from the fact that the data set used may have a number of serious disconcordant observations, not just the single obvious one. In summary, greater power in detection of significant treatment effects was accorded when using iterative reweighted least-squares compared to OLS estimates when influential observations are present.

Nonrandom Assignment of Subjects to Treatment Groups

In Chapter 1 it was mentioned that randomization into treatment groups is the basis for controlled clinical trials, i.e., that every subject has an equal chance of being assigned to a particular treatment group. Dalton and Overall (1977) argue, and present some Monte Carlo simulation supporting their argument, that nonrandom assignment of subjects to groups in an ANCOVA is not only valid but offers some advantages over random assignment to groups. Their assignment algorithm, called the alternate ranks design (ARD), assigns subjects to treatment groups thereby removing any possibility for a correlation between treatment and pretest. The method requires that all subjects be ranked in descending order on their pretest scores. Assuming that there are only two treatment groups, the highest score is assigned to A, the second and third scores are assigned to B, the fourth and fifth scores are assigned to A, the next pair to B, etc. until all subjects have been assigned to a treatment group (Table 5.6). Dalton and Overall (1977) point out that "systematic nonrandom assignment based on the ranked values of the observed pretest scores has the effect of equating not only group means but the entire pretest distributions within the treatment groups."

TABLE 5.4

PRETEST-POSTTEST DATA WITH OUTLIER

Group 1		Group 2	
Pretest	Posttest	Pretest	Posttest
72	74	91	112
74	76	75	106
76	84	74	104
70	67	83	96
71	79	81	**19**
80	72	72	99
79	69	75	109
77	81	77	98
69	60	81	106
82	72		

TABLE 5.5

**ANALYSIS OF TABLE 5.4 USING ORDINARY
LEAST-SQUARES ANCOVA AND ITERATIVELY
REWEIGHTED LEAST-SQUARES ANCOVA**

Method	Source	Sum of Squares	DF	Mean Square	F-Ratio	Prob>F
OLS*	Regression	1.4	1	1.4	0.0	0.956
	Treatment	1845.7	1	1845.7	4.2	0.057
	Error	7057.0	16	441.1		
	Total	9134.1	18			
* outlier still contained in data set						
OLS**	Regression	58.5	1	58.5	1.4	0.251
	Treatment	3359.1	1	3359.1	81.9	0.001
	Error	615.4	15	41.0		
	Total	4767.8	17			
** outlier deleted from analysis						
Huber	Regression	1.1	1	1.1	0.1	0.796
	Treatment	1286.9	1	1286.9	84.6	0.001
	Error	243.3	16	15.2		
	Total	1541.4	18			
bisquare	Regression	1.4	1	1.4	0.6	0.470
	Treatment	559.9	1	559.9	245.3	0.001
	Error	13.7	6	2.28		
	Total	648.4	9			

TABLE 5.6

**EXAMPLE OF A BLOCKED ALTERNATIVE RANKS
DESIGN WITH TWO TREATMENT GROUPS**

Block	Subject	Pretest	Treatment
1	1	90	A
1	2	89	B
1	3	88	B
1	4	84	A
	....		
1	30	62	B
2	31	92	A
2	32	85	B
2	33	81	B
	...		
2	60	65	B
3	61	87	A
3	62	87	B
3	63	85	B
	...		
3	90	61	A
4	91	97	A
4	92	95	B
4	93	90	B
	...		
4	120	64	A

The authors used Monte Carlo simulation to examine whether this method would lead to biased estimates of the true treatment effect and of the significance of these effects. On the basis of their results, they concluded that alternate ranks assignment did not lead to biased estimates of treatment effects and that the precision of these estimates was slightly less than those obtained from random assignment to groups. They also concluded that the power of the F-test in ANCOVA was not changed when subjects were nonrandomly assigned to treatment groups.

It was not the aim of the authors to show that the alternate ranks design was superior to random assignment to treatment groups. Nevertheless, their method appears to have numerous advantages over randomization. Because their algorithm is fixed, assignment to treatment groups removes any prejudices or biases the experimenter may have in assigning subjects to treatment groups. The authors also suggest that "violation of (the) assumptions (in ANCOVA) is less likely using the alternate ranks procedure than it is when subjects are randomly assigned to groups."

Because the alternate ranks design equates the groups in regard to the pretest score variance, it is less unlikely that the design will violate the assumption of homogeneity of variance. Similarly, because heterogeneity of regression coefficients is often the result of baseline incomparability and because the alternate ranks design equates the baseline measures between groups, the alternate ranks design may protect against violating the assumption of homogeneity of regression coefficients. The use of ARD appears to offer many advantages over randomization but is flawed in the sense that reviewers expect to see randomized treatment assignment. Only with better education on the advantages of this method, through more research on its properties and its use in actual studies, will this reviewer bias be overcome.

Summary

- Analysis of covariance is a popular method used to analyze pretest-posttest data.
- ANCOVA implicitly takes into account regression towards the mean.
- Often the assumptions of ANCOVA are violated. The most common violations include:
    1. Use of a parallel line ANCOVA model when the regression slopes between groups are not equal.
    2. Use of ANCOVA when the regression slope is not linear.
    3. When the pretest score is measured with significant error.
    4. When outliers are present which significantly influence the estimation of the regression coefficient.
- Glass, Peckham, and Sanders (1972) present a thorough overview of the consequences of violating the assumptions of ANOVA and ANCOVA. The effect of violating these assumptions on the power and Type I error rate of a statistical test will depend on the severity of the violation.
- One solution to most violations has been the use of some type of rank transform, either to the posttest or to both the pretest and posttest.
    ◊ Rank-transform ANCOVA has similar power and Type I error rate to parametric ANCOVA when the assumptions of parametric ANCOVA are met and it has superior power when the assumptions of parametric ANCOVA are violated.
    ◊ The use of nonparametric, rank-transformed ANCOVA in the analysis of pretest-posttest data should be highly encouraged.
    ◊ One caveat to the rank transformation is that all Monte Carlo simulations which tout its usage have been based on the results of a singular violation of the ANCOVA assumptions. There has been no systematic research examining what to do when multiple irregularities occur. In this case, one may have to choose what is the lesser of the two evils.

- Many argue that the main advantage of ANCOVA over other tests is statistical power and that its use should be encouraged over other tests.
- Alternate ranks treatment assignment appears to have many advantages over randomization to treatments.

CHAPTER 6

BLOCKING TECHNIQUES

Another method to control for the pretest is to either assign subjects to blocks prior to the assignment of a subject to a treatment and then assign treatments to subjects within blocks or to assign subjects to blocks after collection of the posttest and use the post-hoc blocks as a factor in the statistical analysis. The former refers to stratification or blocking, whereas the latter is referred to as post-hoc blocking. Another alternative is assigning subjects to treatments based on an alternate ranks design, but that was already discussed in the previous chapter. Both stratification and post-hoc blocking treat blocks (pretest scores) as a nuisance variable in the experimental design because we are not interested in the nuisance variable per se. We are, however, interested in the posttest because the posttest contains the information needed to determine if a significant difference exists among treatment groups.

Using Stratification to Control for the Pretest

Up to this point, it has been assumed that subjects are randomly assigned to treatments regardless of their pretest scores (except for the discussion on the alternate ranks design). Another method of assigning subjects to treatments is by stratification, which is a restricted randomization process such that treatment groups are made comparable with regards to the pretest state prior to treatment assignment. Often this is accomplished by grouping subjects into blocks and then randomly assigning treatments to each subject within blocks. For example, suppose there are 24 subjects in a two treatment experimental design ranked according to their pretest scores. One block may consist of the first eight subjects with the highest pretest scores, the second block may consist of the next eight subjects, and the last block may consist of the remaining subjects with the lowest pretest scores. It is assumed that subjects within blocks are more homogenous than subjects in other blocks. Treatment assignment in stratified designs is done by randomly assigning the treatments to each subject within blocks. Alternate ranks design (see Chapter 5) is actually a modified stratification method, whereby subjects are grouped into blocks and then, instead of randomly assigning treatments to subject, treatments are systematically applied to subjects within blocks.

By definition, a block consists of a group of experimental units, i.e., subjects, such that the variability within each block is less than the variability among all the blocks. There must be at least two blocks for inclusion in a statistical analysis and three conditions must be met in order for blocking to be valid

1.   There must be at least one treatment with at least two levels and one nuisance variable that can be classified into at least two

levels. For pretest-posttest designs, the nuisance variable is the pretest score.

2. The subjects of interest must be able to form q blocks such that the variability within each block is less than the variability among all subjects in different blocks, i.e., the variability within blocks must be less than the variability among blocks.

3. The levels of the treatment of interest must be applied randomly to subjects within each block.

When all three criteria are met then a randomized block design may be used to analyze the data.

Once the blocks are formed, the analysis of a randomized block design proceeds as an analysis of variance with the linear model being

$$Y_{ijk} = \mu + \alpha_j + \tau_i + e_{ijk} \quad (i = 1, ...p; j = 1, ...q; k = 1, 2, ...n) \tag{6.1}$$

where Y_{ijk} is the posttest score for the kth subject in the jth block and ith treatment, μ is the population mean, α_j is the jth block effect subject to the restriction that the block effects are independent and normally distributed with mean 0 and variance σ_α^2, τ_i is the ith treatment effect subject to the restriction

$$\sum_{i=1}^{p} \tau_i = 0,$$ and e_{ijk} is the random error associated with Y_{ijk}. The errors are

assumed independent and normally distributed with mean 0 and variance σ_e^2 and are independent of α_j and τ_i. As can be seen from the restrictions in the linear model, blocks are treated as random effects.

This model assumes that there is no interaction between blocks and treatments, i.e., that subjects do not differentially respond to treatments. Suppose that subjects who score high on the pretest (and are grouped into a particular block) tend to score higher with a particular treatment than other subjects with lower pretest scores given the same treatment. The two groups will have a different trend in their treatment response and thus violate the assumption of no interaction. The appropriate model where blocks and treatment interact is

$$Y_{ijk} = \mu + \alpha_j + \tau_i + (\alpha\tau)_{ij} + e_{ijk} \quad (i = 1, ...p; j = 1, ...q; k = 1, 2, ...n) \tag{6.2}$$

where $(\alpha\tau)_{ij}$ is the ijth interaction effect for the jth block and treatment level i with mean 0 and variance $\sigma_{(\alpha\tau)}^2$. In the case where the treatment by blocks interaction is significant, it is recommended that PROC MIXED in SAS be used instead of PROC GLM because PROC MIXED computes the exact expected mean square, whereas PROC GLM does not. It has been shown that when the treatment by block interaction term is statistically significant, this is equivalent to violation of the assumption of homogeneity of regression coefficients in ANCOVA.

To determine if a block by treatment interaction results in a "better" model, a lack of fit test may be done. The test statistic is:

$$F = \frac{\left[SSE(R) - SSE(F)\right] \Big/ \left[df_R - df_F\right]}{SSE(F) \Big/ df_F} \qquad (6.3)$$

where SSE(R) is the residual sum of squares in the additive linear model in Eq. (6.1), SSE(F) is the residual sum of squares from the non-additive model in Eq. (6.2), df_R and df_F is the residual degrees of freedom in the additive and non-additive models, respectively. If F is greater than a critical value based on the F-distribution with (df_R - df_F) and df_F degrees of freedom, the null hypothesis is rejected and the non-additive model is used.

One question that arises in choosing to block subjects by their pretest scores is "how many blocks should be used?" Feldt (1958), in a classic paper on blocking in pretest-posttest designs, published a table which may be useful to answer this question. The researcher, however, must know *a priori* or at least have a good estimate of the correlation between pretest and posttest to make use of the table. This table is reprinted in Table 6.1. To use the table the researcher needs to know the total sample size, the correlation between pretest and posttest, and the number of treatment groups. The table is designed for two or five treatment groups. When the number of groups is between two and five, interpolation is necessary. In developing the table, Feldt made two assumptions that still must be met for the table to be valid. One assumption was that the between-groups regression coefficients between pretest and posttest were equal (similar to that seen with ANCOVA). The other was that the variance between groups was equal (homoscedasticity). There have been no studies examining if the optimality of these block sizes still holds when the assumptions are violated. It would wise for the researcher to use caution in this case.

Feldt (1958) also compared the power of analysis of covariance, blocking, and analysis of difference scores and found that the power of the different methods depended on the strength of the correlation between pretest and posttest. When the correlation was less than 0.4, blocking resulted in greater power than ANCOVA, whereas when the power was greater than 0.6, ANCOVA had greater power. When the correlation was between 0.4 and 0.6, both methods had approximately equal power. Also, when the correlation was less than 0.2, neither method had greater power than analysis of variance on the posttest scores. Analysis of difference scores consistently had less power than the other methods regardless of the correlation between pretest and posttest.

TABLE 6.1

**OPTIMAL NUMBER OF BLOCKS IN A RANDOMIZED
BLOCK DESIGN PRESENTED BY FELDT (1958)**

ρ	k	Total Sample Size					
		20	30	50	70	100	150
0.2	2	2	3	4	5	7	9
	5	1	2	2	3	4	6
0.4	2	3	4	6	9	13	17
	5	2	3	4	5	7	10
0.6	2	4	6	9	13	17	25
	5	2	3	5	7	9	14
0.8	2	5	7	12	17	23	25
	5	2	3	5	7	10	15
Courtesy of Feldt, L.S., A comparison of the precision of three experimental designs employing a concomitant variable, *Psychometrika*, 23, 335, 1958. With permission.							

Feldt (1958) concluded that when the sample size was sufficiently large, blocking should be preferred over ANCOVA in most educational and psychological studies because rarely is the correlation between pretest and posttest large enough to take advantage of the greater power of ANCOVA. Also, ANCOVA has more assumptions, which might be more difficult to meet, than blocking. Feldt (1958) states that "dependence on the accuracy of the assumed regression model (in ANCOVA) constitutes a severe restriction on the usefulness of covariance techniques. The absence of any regression assumptions in the (randomized block) design...represents a considerable argument in its favor, especially in instances (where) the number of degrees of freedom are fairly large." However, when the sample size is small ANCOVA should be preferred because small sample sizes do not lend themselves readily to blocked experimental design.

Feldt makes some good points, but overstates the case for using blocking techniques over ANCOVA. First, contrary to his statement above, randomized block designs do have regression assumptions, although those assumptions are less restrictive than with ANCOVA. Second, a high correlation (greater than 0.6) between pretest and posttest is more common than Feldt thinks. In choosing between blocking and ANCOVA, verification of the assumptions each method makes should be done before making any decisions based on which method has greater statistical power because if the assumptions of the method are violated, the validity of the method's conclusions may be in question.

Post-Hoc Stratification

As we have seen, one method to control for the influence of the pretest score is to use a statistical control technique, such as ANCOVA, to remove potential biases in an experiment. We have also just seen that subjects can be assigned to blocks prior to administration of the treatment intervention and then treatments assigned to subjects within blocks, either randomly or systematically. Another type of blocking procedure is to assign subjects into blocks on the basis of their pretest scores after completion of the study. This procedure is known as post-hoc blocking and might be useful when pretest scores are not available at the time of treatment assignment to subjects.

The most difficult question to answer in post-hoc blocking is "how many blocks should be used in the experimental design?" Gill (1984) suggests that the most efficient design is one where the number of subjects per block is equal to the number of treatments. Thus the number of blocks, q,

$$q = \left(\frac{n}{p} \right) \tag{6.4}$$

where n is the total number of subjects, p is the number of treatments, and q is rounded to the nearest integer.

Alternatively, it may be argued that the best choice for the number of blocks is that design which provides the best estimate of the true treatment effect. Unfortunately this can never be known. Instead, we may suggest that the optimal choice for the number of blocks is the number of blocks which leads to the most efficient experimental design. Traditionally, efficiency is compared to the observed mean square error completely ignoring blocking. Therefore, the relative efficiency of the randomized block design may be calculated as

$$RE = \frac{(df_b + 1)(df_{crd} + 3)}{(df_b + 3)(df_{crd} + 1)} \cdot \frac{\sigma_b^2}{\sigma_{crd}^2} \tag{6.5}$$

where df is the error term degrees of freedom, σ^2 is the mean square error term and the subscripts 'b' and 'crd' refer to the blocked design and the completely randomized design, respectively. The ratio of degrees of freedom in Eq. (6.5) is a correction term that reflects the different degrees of freedom between the models. Relative efficiency may be viewed as the number of replications collected on the same subject that must be made if the completely randomized design is used instead of the randomized block design. Alternatively, relative efficiency may be viewed as the ratio of the number of subjects required by the unblocked design to achieve the same power as the blocked design under similar testing conditions. By varying the number of blocks from 2 to q (q <= total number of observations), an iterative estimate of the significance of the block effect may be obtained from which one may decide how many post-hoc blocks to use.

As an example of post-hoc blocking, consider the sexual harassment data. Data were collected on 96 subjects, the subjects were randomized to treatment groups, the treatment applied, and the posttest measured. Each subject was ranked according to pretest score in ascending order and then assigned to a particular block. To make things simple, the number of blocks used were only those blocks for whom an equal number of subjects could be assigned. Thus the number of blocks studied was {2, 3, 4, 6, 8, 12, 16, 24, 32, and 48}. The mean square error of each design was calculated for each of the blocking scenarios and the relative efficiency calculated using Eq. (6.5). A plot of the relative efficiency and mean square error is shown as a function of the block size in Figure 6.1. As can be seen, mean square error drops when the number of blocks equals two and then remains relatively constant. In contrast, relative efficiency peaks at two, then slowly declines to baseline in a somewhat exponential manner thereafter. It would appear that the optimal number of blocks was two since that was the point of the smallest mean square error and highest relative efficiency. Interestingly, this differs somewhat from Gill's recommendation in that the number of blocks should be three.

One disadvantage that may be encountered with post-hoc blocking is the potential for unbalanced designs; there may be an unequal number of subjects in each block or completely empty blocks (Bonett, 1982; Maxwell, Delaney, and Dill, 1984). In the past, unbalanced designs were an issue because statistical packages, such as PROC GLM in SAS, had estimability problems when unbalanced data were present. However, with the development of PROC MIXED in SAS, estimability problems are not an issue as long as there are no totally empty fixed effect cells.

Although post-hoc blocking may be useful, its use should be discouraged at this time for a number of reasons. First, the conditions under which post-hoc blocking may be used are unstudied. Very little systematic research has been done examining the utility of post-hoc blocking in the analysis of pretest-posttest data. In one study, however, Maxwell, Delaney, and Dill (1984) showed that for any combination of correlation and sample size, ANCOVA was more powerful than post-hoc blocking with the greatest increase in precision seen when the correlation was 0.50 or 0.67. Second, post-hoc blocking, with the number of blocks dependent on the data, appears to be suspicious to many - almost like analyzing the data until finding the result the researcher hopes for. Until further research is done, statistical considerations favor ANCOVA over post-hoc blocking.

Summary

- Blocking is less restrictive than ANCOVA because it does not depend on a linear relationship between the pretest and posttest scores.

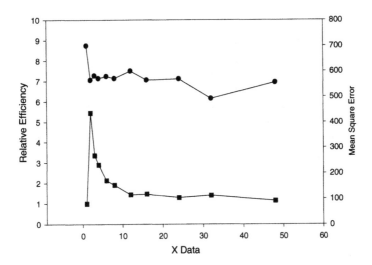

Figure 6.1: Relative efficiency of post-hoc blocking (solid square) and mean square error (solid circle) as a function of block size. Maximum efficiency is reached using two blocks, whereas minimum mean square error remains relatively constant throughout.

- The randomized block design is appropriate even if the relationship between pretest and posttest scores is nonlinear in nature.
- The choice of the number of blocks to use is difficult.
  - ◊ One suggestion is to use the same number of blocks as the number of treatment groups.
  - ◊ Another is to choose the number of blocks based on the correlation between pretest and posttest.
  - ◊ Another advocates iterating the number of blocks post-hoc and finding that number which provides the maximal efficiency.
  - ◊ Until further research is done, Feldt's guidelines on choosing the number of blocks should be followed.
- Maxwell, Delaney, and Dill (1984) provide the following guidelines in choosing ANCOVA over blocking.
  - ◊ Are the pretest scores available prior to assignment of subjects to treatments? If so, power is often increased when subjects are assigned to treatments according to their pretest scores *a priori* (either as blocks or alternate ranks).

◊ Is the relationship between pretest and posttest linear? If so, then ANCOVA is the recommended method of analysis. If not, then either ANCOVA with higher order polynomials or blocking may be used.

◊ In contrast to most others, Maxwell, Delaney, and Dill (1984) argues that choosing a method based on the correlation between pretest and posttest is irrelevant. The only reason one may want to consider the correlation is in deciding if it is sufficiently large ($\rho > 0.3$) to compensate for the loss of degrees of freedom in utilizing a covariate in the analysis.

• The efficiency of a randomized block design is dependent on the number of blocks chosen and its use is preferable to ANCOVA when the sole goal of the experimenter is to reduce experimental error, rather than estimate treatment effects.

CHAPTER 7

REPEATED MEASURES ANALYSIS OF VARIANCE

One method, which may appear to be commonsense, is to analyze pretest-posttest data using a repeated measures analysis of variance. After all, in a repeated measures experimental design, multiple measurements are made on the same subject and the test for treatment effects is based not on experimental error, but on within subject error. We shall see however that this approach is valid only when multiple posttest measurements are made on the same subject. When applied to a single pretest and single posttest, the output from a repeated measures analysis of variance has a unique interpretation.

Using Repeated Measures ANOVA for Analysis of Pretest-Posttest Data

In a repeated measures experimental design, the basic experimental unit is the subject. Each subject is treated as a block that is repeatedly measured over time. On face value, pretest-posttest designs appear to represent a special case of the repeated measures design, wherein only two testing occasions are done on each subject. The linear model for a basic repeated measures design where there is only one between subjects effect is

$$Y_{ijk} = \mu + \alpha_i + \tau_{j(i)} + \beta_k + \alpha\beta_{ik} + e_{ijk} \qquad (7.1)$$

where $i = 1, 2, ... p$ (Treatment 1, Treatment 2, ... p), $j = 1, 2, ... n$ (subject), $k = 1, 2$ (measurement = 1, measurement = 2), Y_{ijk} is the kth measurement made in the jth subject assigned to the ith treatment group, α_i is the ith treatment group effect, $\tau_{j(i)}$ is the jth subject effect nested in the ith treatment group, β_k is the kth time effect, $\alpha\beta_{ik}$ is the treatment by time interaction effect, and e_{ijk} is random error. In order to solve Eq. (7.1), certain assumptions must be made, i.e., $\sum_{i=1}^{p} \alpha_i = 0$, $\sum_{k=1}^{2} \beta_k = 0$, etc. The reader is referred to Neter et al. (1996) or Kirk (1982) for further details on the solution to the repeated measures linear model. The variance of Eq. (7.1) is

$$Var(Y_{ijk}) = Var(\tau) + Var(e) \qquad (7.2)$$

which states that the variance of the observations is equal to the sum of between subject and residual error variance. Eq. (7.2) is the formula that has been repeatedly used herein to define the variance of pretest and posttest observations. On face value, repeated measures appear to be an appropriate method for the analysis of pretest-posttest data. However, as Huck and McLean (1975) have pointed out, ANOVA "can be misinterpreted because the linear...model [in Eq. (7.1)] is not entirely correct. Since the pretest scores are by definition, collected prior to exposure of subjects to the treatment effect(s), it is impossible for the treatment effects (i.e., the α's) or an alleged interaction effect (i.e., the $\alpha\beta$'s) to affect any of the pretest scores."

The model presented by Eq. (7.1) assumes that subjects are randomized to a treatment group, the treatment intervention is applied, and then the measurements are repeatedly collected on each subject. In other words, Eq. (7.1) assumes that all measurements are made **after** imposition of the treatment intervention. In a pretest-posttest design, subjects are measured prior to assignment to a treatment group, randomized to a treatment group, the treatment intervention is applied, and then an additional measurement is collected on each subject. Thus the basic repeated measures linear model is inappropriate for a pretest-posttest design. The correct linear model for the pretest scores is given by

$$Y_{j1} = \mu + \tau_j + e_{j1} \tag{7.3}$$

where the dot notation denotes that subjects are not assigned to groups yet. While the posttest scores are affected by the treatment effect they cannot be affected by the interaction effect. Therefore, the appropriate linear model for the posttest scores is

$$Y_{ij2} = \mu + \alpha_i + \tau_{j(i)} + \beta_2 + e_{ij2} . \tag{7.4}$$

Since the treatment affects only the posttest, and not the pretest, the F-test produced from a repeated measures analysis will be biased in the estimation of the treatment effect. The expected mean square for the F-test for treatment in a repeated measures analysis of variance for pretest-posttest data when the assumptions and experimental design are correct is given by

$$\frac{E(\text{Treatment})}{E(\text{Subjects within Treatment})} = \frac{\sigma_e^2 + 2\sigma_\tau^2 + 2n\sigma_\alpha^2}{\sigma_e^2 + 2\sigma_\tau^2} . \tag{7.5}$$

If a repeated measures analysis of variance is done on pretest-posttest data, the expected mean square for the F-test for treatment is given by

$$\frac{E(\text{Treatment})}{E(\text{Subjects within Treatment})} = \frac{\sigma_e^2 + 2\sigma_\tau^2 + \frac{n}{2}\sigma_\alpha^2}{\sigma_e^2 + 2\sigma_\tau^2} . \tag{7.6}$$

Because the third term in the numerator of Eq. (7.6) is only one-fourth as large as that in Eq. (7.5), the F-test produced by Eq. (7.6) will be too conservative, thereby increasing the probability of committing a Type II error (failure to reject the null hypothesis when it is true). Thus the traditional repeated measures analysis of variance applied to pretest posttest data will lead to an invalid estimate of the true treatment effect.

Surprisingly, one can still obtain an unbiased estimate of the treatment effect from a repeated measures analysis of variance, not from the F-test of the treatment effect, but from the F-test for the treatment by time interaction (Brogan and Kutner, 1980)! Amazingly, the F-test of the interaction effect and the F-test from an analysis of difference scores will always be the same. As an example consider the data presented by Asnis, Lemus, and Halbreich (1986). Thirteen (13) subjects with endogenous depression and 20 normal controls participated in the study. Each subject's fasting baseline cortisol concentration was measured in triplicate at 9 a.m. The mean of the three values was used as a baseline. Each subject was then administered 75 mg of desipramine, a tricyclic antidepressant. Blood was collected every 15 min for 120 min and the average cortisol level was determined. The data are plotted in the top of Figure 7.1 and the ANOVA table for the analysis of variance is presented in Table 7.1. Looking at only the treatment F-test, one would conclude that there was no difference between groups in cortisol levels (p = 0.9059). This result is in contrast to the analysis of difference scores (bottom of Figure 7.1). The paired samples t-test of the difference scores was -2.45 (p = 0.0201). Using this test statistic, the researcher would conclude that there was a difference between treatments. Amazingly, the results of the t-test are exactly equal to the square root of the treatment by time interaction F-test ($\sqrt{6.00} = 2.45$).

How can the treatment by time interaction give the appropriate estimate of treatment effect when a treatment by time interaction effect is not even included in Eqs. (7.3) and (7.4)? Because in the case of pretest-posttest data, the treatment by time interaction estimates the square of the average difference between the pretest and posttest summed across groups divided by a residual error term, which is equivalent to a one-way analysis of variance on difference scores. So then what exactly does the treatment F-test estimate? The treatment F-test tests the hypothesis that the sum of the pretest and posttest measurements are equal among groups. Going back to the example, the absolute value of the t-test comparing the sum of the pretest and posttest cortisol levels was -0.119 (df = 31, p = 0.9059), which is equal to the square root of the treatment F-test ($\sqrt{0.0142} = 0.119$).

All the information contained from an analysis of difference scores is contained within a repeated measures analysis of variance. However, repeated measures analysis of variance has a greater number of assumptions than difference scores analysis. In particular, repeated measures analysis of variance has all the assumptions the usual analysis of variance has the error term is distributed as a normal random variate with mean 0 with homogenous variance, but also assumes a particular variance/covariance structure between the observations which must be specified in the linear model. PROC GLM in SAS is quite restrictive in the variance/covariance matrix that can be used; the only variance/covariance matrix that can be used is one where the pretest and posttest have equal variance. PROC MIXED is more flexible allowing pretest and posttest to have different variances. The covariance structure in PROC MIXED is specified in the REPEATED statement by using the TYPE= command.

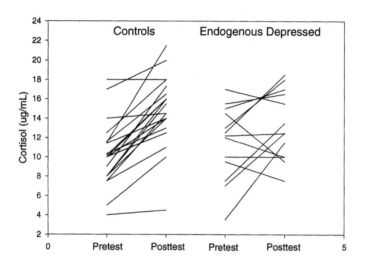

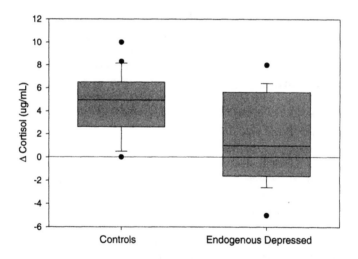

Figure 7.1: Cortisol levels for 13 patients with endogenous depression and 20 normal controls plotted as a tilted-line segment plot (top) and difference scores for patients and controls plotted as a whisker plot (bottom). Data reprinted from Asnis, G.M., Lemus, C.Z., and Halbreich, U., The desipramine cortisol test–a selective noradrenergic challenge (relationship to other cortisol tests in depressives and normals), *Psychopharmacology Bulletin*, 22, 571-578, 1986. With permission.

TABLE 7.1

**REPEATED MEASURES ANOVA TABLE FOR THE DATA
OF ASNIS, LEMUS, AND HALBREICH (1986)**

Source	DF	Sum of Squares	Mean Square	F-ratio	Prob > F
Treatment	1	0.30	0.30	0.01	0.91
Subject(Treatment)	31	650.94	21.00	3.87	<0.01
Time	1	163.94	163.94	30.25	<0.01
Treatment by time	1	32.54	32.54	6.00	0.02
Error	31	168.01	5.42		
Total	65	1057.43			

In summary, using repeated measures analysis of variance to analyze the data when the pretest is collected prior to administration of the treatment intervention is inappropriate because the underlying model is misspecified. Fortunately, however, one can still test the null hypothesis of no treatment effect using the output of a repeated measures analysis of variance by examination of the treatment by time interaction F-test, not the treatment test. It is the author's opinion, however, that repeated measures analysis of variance not be used, not for any technical reason, but because it is difficult to explain to a person unfamiliar with the nuances of the repeated measures analysis applied to pretest-posttest data why the appropriate F-test for comparing treatments is not the treatment F-test. A naive reviewer of one's work would think that he was statistically ignorant and reject his conclusions out of hand.

Regression Towards the Mean with Multiple Posttest Measurements

When more than one posttest measurement is made on the same subject, a true repeated measures analysis of variance applies, as one would normally expect. It is beyond the scope of this book to cover this topic in detail because entire books have already been devoted to this topic. The reader is referred to Littell et al. (1996) or Diggle, Liang, and Zeger (1995) for greater detail. One of the most basic concepts that was developed early on in this book was regression towards the mean, i.e., that extreme pretest scores (compared to the population average) will tend to regress towards the population mean on the second measurement. To see how regression towards

the mean extends from testing on two occasions to multiple occasions one must examine the stochastic nature of multiple testing situations. Recall from previous chapters that the expected value of the posttest given the pretest is given by

$$E(Y|X = x) = \mu_Y + \frac{\rho(x - \mu_x)\sigma_Y}{\sigma_X}. \tag{7.7}$$

To simplify the concepts, transform the data to standardized values such that $\sigma_Y = \sigma_X = 1$ and $\mu_X = \mu_Y = 0$, such that $E(Y|X = x) = \rho x$. Also, let $Y_1, Y_2, \ldots$ Y_n be the first, second, and nth posttest measurements, respectively. By extension of Eq. (7.7) one might naively predict that for multiple testing occasions the expected value of the ith measurement is dependent on previous measurements. Thus

$$E(Y_1|X=x) = corr(X, Y_1)x \tag{7.8}$$

$$E(Y_2|Y_1=y) = corr(Y_1, Y_2)y \tag{7.9}$$

$$E(Y_3|Y_2=z) = corr(Y_2, Y_3)z \tag{7.10}$$

and so forth. In other words, the ith measurement is correlated with the (ith-1) measurement. Although these equations are technically correct, they are misleading because they conclude that as the sequence progresses each observed value will get closer and closer to the mean of 0, i.e., the impact of regression towards the mean becomes greater and greater. To see this, substitute Eqs. (7.8) and (7.9) into Eq. (7.10). Hence,

$$E(Y_3|Y_2 = z) = corr(x,Y_1)*corr(Y_1,Y_2)*corr(Y_2,Y_3)*z.$$

If $corr(Y_i, Y_{i+1}) = 0.5$ and $x = 2$, then,
$$Y_1 = 1, Y_2 = 0.5, Y_3 = 0.25, \text{etc.},$$
until $Y_p = 0$. As Nesselroade, Stigler, and Baltes (1980) point out, this sequencing procedure is incorrect because it appears as though regression towards the mean is "steady and unrelenting" which clearly does not occur.

The correct expected sequence is dependent only on the nature of the correlation between the first and the ith observation, not on the correlation of the intermediate observations. The correct expected sequence then becomes

$$E(Y_1|X = x) = corr(X, Y_1)x$$

$$E(Y_2|X = x) = corr(X, Y_2)$$

$$E(Y_3|X = x) = corr(X, Y_3)x$$

etc. Thus the pattern of autocorrelations between pretest and the ith posttest, $corr(X, Y_i)$, will determine the pattern of the expected sequence. It can be shown that for more than two observations, the observed pattern is not always one of regression towards the mean. For example, if constant correlation between X and Y_i is assumed, the expected sequence becomes $\rho x, \rho x, \ldots \rho x$. In this case, regression towards the mean is said to have dissipated after the second occasion. In general, whenever the correlation between pretest and the ith posttest is either constant or a decreasing function over time, then regression towards the mean occurs between the pretest and first posttest measurement, but is dissipated thereafter. Nesselroade, Stigler, and Baltes

(1980) present cases where other autocorrelation patterns occur between observations and the impact they have on regression towards the mean.

One practical outcome of this work is its impact on experimental design. If regression towards the mean declines or reaches a steady state during a sequence of observations, then practicing a test before administration of a treatment should reduce the confounding effect regression towards the mean has on estimation of the treatment effect. This validates the guideline provided by McDonald, Mazzuca, and McCabe (1983) that the average multiple pretest scores should be used to minimize the impact of regression towards the mean.

Using Repeated Measures ANOVA for Analysis of Pretest-Posttest Data with Multiple Posttest Measurements

Given that multiple posttest measurements made on the same subject are true repeated measures experimental designs and that the F-tests obtained from this analysis are correct, the question becomes "what to do with pretest data?" Obviously all the same methods that have been presented up until now are still valid. One method is to compute the difference scores or percent change scores for each subject and to analyze the difference or change scores using an analysis of variance with time, treatment, and treatment by time interaction as fixed effect factors. These methods are perfectly valid, but the restrictions and cautions that apply in the single posttest measurement case still apply with the multiple posttest case.

Probably the best method to analyze multiple posttest measurement data when baseline data is collected on each subject is to treat the data as a mixed effect linear model with covariance analysis. This approach is an extension of the last chapter on ANCOVA. In the case where a treatment intervention is applied to groups of subjects the linear model is

$$Y_{ijk} = \mu + \alpha_i + \tau_{j(i)} + \beta_k + \alpha\beta_{ik} + \gamma(X_j - \mu) + e_{ijk} \qquad (7.11)$$

where all factors are treated as in Eq. (7.1) and $\gamma(X_j - \mu)$ is the dependence of the posttest score on the pretest score. An excellent example of this approach is given in Wolfinger (1997) in the repeated measures analysis of systolic blood pressure data from a clinical trial studying the effect of various medications on hypertension. The reader is also referred to Littell et al. (1996) for further details on the SAS code and the use of PROC MIXED to analyze this type of data.

Analysis of Repeated Measures Using Summary Measures

Pocock (1983) and Frison and Pocock (1992) proposed that repeated measures data be analyzed by analysis of summary measures instead of raw data. The idea is that all posttest measurements within a subject can be collapsed into a single, univariate composite score that is still correlated with the pretest score and can be used in a linear mixed effects model with pretest scores as the covariate. Summary measures remove the effect of the time component by manipulation of the posttest scores into a univariate summary statistic. They have a number of other advantages in addition to simplification of the problem at hand. They are less sensitive to the influence of missing data and are easier to both understand and analyze. There are many different summary measures, some of which we are already familiar with, change scores and percent change scores being just two of them. Other summary measures include the average of all posttest measures, maximal or minimal posttest measurements, maximal or minimal change or percent change score, and area under the curve.

For example, in the development of a new non-cardiac-related drug, it is suggested by regulatory agencies that the drug be examined for its effects on cardiac repolarization. This suggestion arose when it was discovered that terfenadine, a widely used prescription anti-histamine, may result in sudden death when combined with erythromycin, an antibiotic. Further examination revealed that terfenadine increases the time for cardiac repolarization as evidenced by an increase in the QTc interval on an electrocardiogram. In general, if at the time of maximal drug effect the QTc interval is greater than some cut-off value, the drug is said to prolong QTc intervals. In keeping with the new regulatory guidelines, suppose a drug company administers its new drug candidate to a group of patients and measures the QTc interval in patients repeatedly over time after drug administration. In this case, the drug company is not interested in the correlation or inter-dependence of observations over time but in the maximal QTc interval observed in a subject. What is of interest is "is there a significant difference between the maximal observed QTc interval in a patient and his baseline drug-free QTc interval." This is an example of using the maximal change from baseline as a summary measure of drug effect. In this case, a multivariate problem is reduced to a single posttest score at the time of maximal effect. Now any of the methods discussed up to this point may be used to analyze the data.

Another example of a summary measure is where the data can be collapsed into an "overall effect measure." For example, drug X was being developed as an antibiotic. A drug that might be co-prescribed with drug X is rifampin, a macrolide antibiotic. However, rifampin has been previously shown to increase the metabolism of many drugs that are metabolized by the liver, such as oral contraceptives, and it was of interest to determine if rifampin would increase the metabolism of drug X. A study was conducted wherein 14 volunteers were given drug X and the concentration of drug X in plasma was measured over time. After a washout period, the volunteers returned, were given rifampin daily for 3 days, and then challenged with drug X again. Plasma concentrations of drug X were measured and compared against the baseline data. The data are shown in Figure 7.2. In this case multiple pretest (without rifampin) and posttest (after rifampin) data were

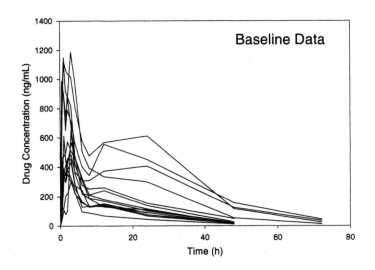

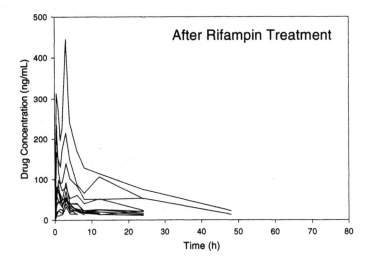

Figure 7.2: Drug X plasma levels over time after oral administration of drug X in 14 volunteers before (top) and after (bottom) rifampin pretreatment.

available. As a simplification of the data, the analyst calculated the area under the curve (AUC) for each subject. This represents the overall degree of drug exposure each subject experiences and can be thought of as the integral of drug concentrations from time zero to the last time point for which data are available on that particular subject. An alternative representation is that AUC is a weighted average where the weights are proportional to the time intervals between measurements. The plasma concentration-time profile for Subject 1 is shown in Figure 7.3. The hatched area is the AUC, which was calculated using the linear trapezoidal rule

$$AUC = \sum_{i=1}^{n-1} 0.5 \cdot (t_{i+1} - t_i)(C_{i+1} + C_i) \qquad (7.12)$$

where n is the number of time points for which data are collected, t is time, and C is plasma concentration, the dependent variable. Note that C can be replaced by any continuous random dependent variable. Figure 7.4 shows the AUC for each subject before and after rifampin treatment. It can be seen that what was originally a multivariate problem has been reduced to a simple pretest-posttest design. Both before and after rifampin administration, AUC was not normally distributed, a not uncommon situation because AUC is usually log-normal in distribution (Lacey et al., 1997). After natural log transformation the data were normally distributed. Ignoring the possible effect the distribution of the data would have on the conclusions, it was decided that a nonparametric method would be used. Wilcoxon's signed rank test was done on the paired difference scores and found to be highly significant (test statistic = -52.5, $p < 0.0001$). It was concluded that rifampin has a significant drug interaction with drug X. Another example of this approach can be found in a study of the interaction between quinidine and nifedipine (Bowles et al., 1993). AUC analyses can be extended to many repeated measures problems when the goal of the study is to find out if there is an overall difference between treatments (Dawson, 1994). However, traditional repeated measures analysis using time as a categorical or continuous variable is useful because it allows for isolation of specific time periods where differences may occur between treatments.

There are a variety of modifications of the AUC summary measure, including the time averaged AUC

$$AUC_{TA} = \sum_{i=1}^{n-1} \frac{0.5 \cdot (t_{i+1} - t_i)(Y_{i+1} + Y_i)}{t_n - t_1} \qquad (7.13)$$

where AUC_{TA} is normalized to the total time interval or time-averaged AUC with baseline correction (AUC_{TA-BC})

$$AUC_{TA-BC} = \sum_{i=1}^{n-1} \frac{0.5 \cdot (t_{i+1} - t_i)(Y_{i+1} + Y_i)}{t_n - t_1} - Y_0 \qquad (7.14)$$

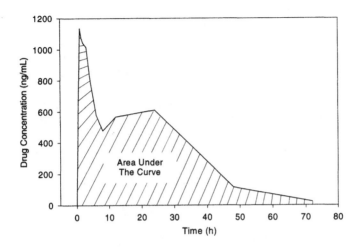

Figure 7.3: Figure demonstrating the concept of area under the curve. Data are plasma drug X concentrations over time for a single subject. The shaded area represents the area under the curve.

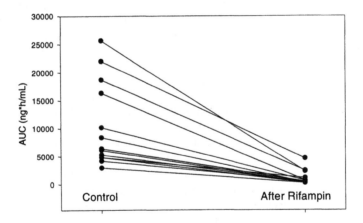

Figure 7.4: Tilted line-segment plot for the area under the curve (AUC) data in Figure 7.2. Rifampin pretreatment showed a clear negative effect on AUC. This plot shows that what was once a multivariate problem (Figure 7.2) can be reduced to a univariate one (Figure 7.4) using an appropriate summary measure.

where Y is the dependent variable and Y_0 is the baseline. Raboud et al. (1996) recommend some form of AUC as summary measure in the analysis of data which use plasma RNA from HIV clinical trials.

Yet another example of a summary measure is to use the average posttest score. This summary statistic assumes that there is no underlying time component and that all variation in observations within a subject is due to random error. An example of this is the assessment of QTc prolongation by terfenadine (Pratt et al., 1996). In this study the dose-response relationship between terfenadine and QTc prolongation was established using a four-period cross-over design. Granted a cross-over design is not a true pretest-posttest design, it does illustrate the use of average posttest scores as a summary measure. In this study, 28 healthy volunteers and 28 patients with stable cardiovascular disease were randomized to receive placebo, 60 mg terfenadine twice daily (BID), 180 mg terfenadine BID, or 300 mg terfenadine BID in one of four treatment periods. Each treatment was applied for 7 days. Within each period, serial electrocardiogram readings were made on days 0, 1, 5, 6, and 7. Mean QTc interval readings were calculated for days 1 and 5 and the results compared. It was shown that terfenadine results in a dose-dependent prolongation of the QTc interval, an indication that terfenadine prolongs cardiac repolarization, in both healthy volunteers and patients with cardiac disease.

Kuhn and DeMasi (1999) used Monte Carlo to study the power of summary measures in a repeated measures study when data were missing at random. Analyzed in the study were the last observed posttest scores, baseline corrected time-averaged AUC, maximum posttest scores, median difference scores, and average minus baseline scores. The median difference score performed uniformly better than other measures, whereas analyzing the last observed posttest scores performed uniformly worst. They conclude that most of the summary metrics studied have certain advantages. The metrics are useful when missing data are present and capture most of the relevant information in a subject's time profile.

Donahue (1997) presented a newly devised nonparametric summary statistic called the W-statistic which can be used as the dependent variable in an ANOVA to examine differences between treatments. Suppose for the sake of argument that each subject has n (n ≥ 1) pretest and m (m ≥ 1) posttest measurements. If each subject's n pretest scores are sorted in ascending order from $X_{(1)}$ to $X_{(n)}$, the W-statistic for that subject is given by

$$W = \sum_{i=1}^{n} W_i \qquad (7.15)$$

where

$$W_i = \begin{cases} 1, & \text{if } Y_i > X_{(n)} \\ -1, & \text{if } Y_i < X_{(1)} \\ 0, & \text{otherwise} \end{cases}.$$

Simply put, W is the number of Y_i's (posttest scores) greater than the maximum of X_i's (pretest scores) minus the number of Y_i's less than the minimum X_i's. For example, if the pretest scores are 3.9 and 3.0 and the posttest scores were: 2.8, 3.7, 4.5, 3.2, 3.2, 1.8, and 3.2, W equals (-1). The W-statistic represents the change in location or location shift which took place between measurement of the pretest and after administration of the treatment intervention. Assuming there is no change in posttest scores compared to pretest scores, the p-value for any individual observed W statistic is given by

$$p(W = w) = p(W = -w) = \sum_{k=0}^{\left[\frac{m-w}{2}\right]} \frac{\binom{n+m-2k-w-2}{n-2}}{\binom{n+m}{2}} \qquad (7.16)$$

where (.) is the binomial operation,

$$\binom{n}{r} = \frac{n!}{r!(n-r)!} \qquad (7.17)$$

where ! represents the factorial function, and the (.) operator atop the summation sign represents the greatest integer, i.e., if (m-w)/2 equals 1.5, then summation commences from 0 to 2. As an example, given that two pretest measurements are made and five posttest measurements are made, what is the probability that w = 3? Substituting these values into Eq. (7.16) gives

$$p(W = 3) = \sum_{k=0}^{\left[\frac{5-3}{2}\right]} \frac{\binom{2+5-2k-3-2}{2-2}}{\binom{2+7}{2}} = 0.0952.$$

Table 7.2 gives the p-values for various values of n, m, and W. Using Monte Carlo simulation comparing the W-statistic in an ANOVA to mean change ANCOVA with baseline pretest mean as the covariate, Donahue (1997) showed that the W-statistic has a Type I error rate near the nominal level α and has power slightly less than parametric ANCOVA using the pretest scores as the covariate. However, when baseline scores were distributed as a non-normal distribution (in this case the Cauchy distribution was used) with heavy tails, ANOVA using the W-statistic had significantly greater power at detecting a treatment difference than did parametric ANCOVA using the

TABLE 7.2

p-VALUES FOR DONAHUE'S (1997) W-STATISTIC
FOR VARIOUS VALUES OF n, m, AND W

n	m	W = 0	1	2	3	4	5	6
2	2	0.3333	0.1667	0.1667				
2	3	0.2000	0.2000	0.1000	0.1000			
2	4	0.2000	0.1333	0.1333	0.0667	0.0667		
2	5	0.1429	0.1429	0.0952	0.0952	0.0476	0.0476	
2	6	0.1429	0.1071	0.1071	0.0714	0.0714	0.0357	0.0357
3	3	0.3000	0.2000	0.1000	0.0500			
3	4	0.2571	0.1714	0.1143	0.0571	0.0286		
3	5	0.2143	0.1607	0.1071	0.0714	0.0357	0.0179	
3	6	0.1905	0.1429	0.1071	0.0714	0.0476	0.0238	0.0119
4	4	0.3143	0.1857	0.1000	0.0429	0.0143		
4	5	0.2698	0.1746	0.1032	0.0556	0.0238	0.0079	
4	6	0.2381	0.1619	0.1048	0.0619	0.0333	0.0143	0.0048

Note: A more extensive table can be found in Donahue (1997). Reprinted from Donahue, R.M.J., A summary statistic for measuring change from baseline, *Journal of Biopharmaceutical Statistics*, 7, 287-289, 1997 by courtesy of Marcel Dekker, Inc.

pretest scores as the covariate. It was concluded that the W-statistic is more resistant to outliers than ANCOVA. However, the W-statistic is still quite new and its validity has not been rigorously challenged. It does appear to be a useful summary measure for repeated measures design and its use should not be discouraged. In summary, many times it is possible to transform what was originally a multivariate problem into a univariate problem by using either AUC, Donahue's W-statistic, maximal effect, or some other transformation. Once transformed the data may be analyzed by methods presented in earlier chapters.

Summary
- Use of the F-test for treatment effect in the analysis of simple pretest-posttest data using a repeated measures results in a biased estimate of the true treatment effect.
- The appropriate F-test in a repeated measures analysis of variance on pretest-posttest data is the treatment by time interaction.
- Multiple posttest measurements after the treatment intervention is a true repeated measures design and can be analyzed by either a traditional

repeated measures analysis of variance with baseline pretest score as a covariate or using summary measures.

- Regression towards the mean still occurs with multiple posttest scores, but the extent of regression towards the mean depends on the pattern of autocorrelation between observations.
- Sometimes when multiple pretest and/or posttest data are collected on an individual it is possible to generate summary statistics, such as area under the curve, Donahue's (1997) W-statistic, or maximal effect, which transforms a multivariate analysis into a univariate one. Once the problem is simplified, any of the tests developed in this book can be used to analyze the data.

CHAPTER 8

CHOOSING A STATISTICAL TEST

Choosing a statistical test to analyze your data is one of the most difficult decisions a researcher makes in the experimental process because choosing the wrong test may lead to the wrong conclusion. If a test is chosen with inherently low statistical power, a Type II error can result, i.e., failure to reject the null hypothesis given that it is true. On the other hand, choosing a test that is sensitive to the validity of its assumptions may result in inaccurate p-values when those assumptions are incorrect. The ideal test is one that is uniformly most powerful and is not sensitive to minor violations of its underlying assumptions. It would be nice to be able to say, always use test X for the analysis of pretest-posttest data. Unfortunately, no such test exists. The tests that have been presented in previous chapters may be best under different conditions depending on the observed data. The rest of this chapter will be devoted to picking the right test for your analysis. It should be pointed out that the choice of statistical test should be stated and defined prior to doing the experiment.

Choosing a Test Based on How the Data will be Presented

Up to now each of the statistical tests have been presented in terms of rejection or acceptance of the null hypothesis. If a set of data were analyzed by every method presented, every test may come to the same conclusion, e.g., rejection of the null hypothesis. Even though all the tests may give equivalent results they will not, however, result in the same interpretation (Wainer, 1991). An example of how each statistical test's interpretation may vary slightly, consider the case where a researcher is interested in determining if men and women recall the same number of words on a memory task after administration of a drug which causes anterograde amnesia. If initial values are ignored (posttest-only analysis), the following descriptive statement may be presented

> On average, men recall 13 words and women recall 8 words after drug administration.

If the baseline score is subtracted from the on-drug values (difference scores analysis), the statement may be modified to

> On average, drug administration decreased the number of words recalled by 12 in men and 9 in women.

If the data were transformed to percentages (percent change analysis), then the following statement may be presented

> On average, drug administration decreased the number of words recalled by 60% in men and 40% in women.

If the final scores were adjusted by their drug-free scores with covariance adjustment (ANCOVA), then the following may be stated

On average, the decrease in the number of words recalled by men was similar to women with the same baseline.

Each of these statements is more or less the same − stating the difference between men and women. But each does so in a unique way that adds a different level of meaning to the interpretation.

Sometimes the statistical test chosen to analyze pretest-posttest data is based on how the data will be presented to the intended audience. It may be easier to present the results of one type of analysis than another. For example, suppose an individual takes the digit-symbol substitution test of the Wechsler Adult Intelligence Scale as a baseline measure of psychomotor performance. After administration of a new experimental drug, that individual's maximum degree of impairment was a difference score of 20. It would be difficult to explain the significance of this difference to a non-psychologist. However, explaining that the drug decreased psychomotor performance by 30% is more easily understood by the layman. For this reason the researcher may decide that analysis of percent change scores will be the primary statistical test.

Keep in mind that although percent change may be the best way to express the data, it may not be the best way to analyze the data because not all tests are equal. Each of the tests may have different Type I error rates and statistical power curves. Under certain conditions one type of analysis may have greater ability to detect a difference among the group means than another type of analysis. Using the analysis that has greater ease of interpretation may result in using a test that does not have sufficient power to detect a difference among groups, possibly resulting in a Type II error. The primary question in analyzing the data should be "which statistical test will give me the greatest ability to detect a difference among groups and will result in p-values that are not invalid due to some violation of the assumptions of the test used?" The rest of the chapter will use Monte Carlo simulation to examine how the statistical tests hold up when the assumptions of the test are violated.

Generation of Bivariate, Normally Distributed Data with a Specified Covariance Structure

The basis of Monte Carlo simulation is the ability to generate random numbers having some predefined mean and variance such that they represent the sampling distribution of the population of interest. In this case, Monte Carlo simulation will need to be able to generate pretest scores with a given mean and variance and posttest scores with a given mean and variance. Both pretest and posttest scores must have a specific probability density to them, e.g., normal, log-normal, etc. Also, the set of pretest-posttest scores must have a specific correlation between them. The general Monte Carlo method is further explained by Mooney (1997).

One method to generate bivariate normally distributed data (X,Y) with mean 0 and variance 1, expressed mathematically as

$$X \sim N(0,1) \text{ and } Y \sim N(0,1),$$

with a specified reliability or correlation (ρ) is as follows: let X and W be independent random normal variables with mean 0 and variance 1, then

$$Y = \rho X + W\sqrt{1-\rho^2} \ . \tag{8.1}$$

This process is repeated n times for the desired sample size. If the computer program that is being used does not generate normal random variates then the method of Box and Muller (1958) can be used. Box and Muller's (1958) method depends on the ability of the computer to generate uniformly distributed random variables, which most computer languages readily do. Their method is as follows: if A and B are uniform random variates on the interval (0, 1) then X and W will be normally distributed with mean 0 and variance 1, where

$$X = \cos(2\pi B) * \sqrt{-2 \cdot \ln(A)} \tag{8.2}$$

$$W = \sin(2\pi B) * \sqrt{-2 \cdot \ln(A)} \tag{8.3}$$

and ln is the natural log transformation of A. These random variables can then be substituted into Eq. (8.1) to generate the bivariate normal random variables (X, Y).

Monte Carlo Simulation When the Assumptions of the Statistical Test are Met

In an attempt to understand the conditions under which the statistical tests that were presented in the previous chapters perform best, Monte Carlo simulations were undertaken to examine the power and Type I error rate of each of the tests under a variety of different statistical distributions and degree of correlation. The first simulation was designed to compare the tests when the assumptions of the statistical tests are met, i.e., the data are normally distributed, the relationship between the pretest and posttest is linear in nature, etc. Thirty subjects were randomized to three groups of ten subjects. Pretest and posttest measurements were collected on each group and designed to have a correlation ranging from 0 to 0.95. Bivariate correlated pretest-posttest scores were generated using the method described previously. These scores represent the values each subject would have received had no treatment intervention been applied between measurements of the pretest and posttest. A constant treatment intervention of a given effect size, called τ, was calculated using Eq. (1.5), and then applied to each subject's posttest score within the appropriate group to simulate the effect his appropriate treatment intervention had on him. One group served as a control group and did not receive a treatment intervention. Subsequently this group did not have a treatment effect added to their posttest scores. One group received the maximum treatment intervention and the other group received one-half the maximum treatment intervention. The value of the treatment effect added to each subject's posttest score was such that effect size, calculated using the method of Cohen (1988).

The test statistics that were used to test for treatment differences were as follows (note that the symbols used in the plots that follow are identified with the treatment)

1. ANOVA on posttest scores only (solid circle).
2. ANOVA on difference scores (open circle).
3. ANOVA on rank difference scores using the Blom transformation (solid upside down triangle).
4. ANOVA on percent change scores (open upside down triangle).
5. ANOVA on log-transformed difference scores (solid square).
6. ANOVA using Brouwers and Mohr (1989) transformation (modified percent change scores) (open square).
7. ANCOVA using posttest scores as the covariate (solid diamond).
8. ANCOVA using ranked pretest scores as the dependent variable and ranked posttest scores as the covariate (open diamond).
9. ANCOVA using ranked normal pretest scores as the dependent variable and ranked normal posttest scores as the covariate (solid triangle).
10. ANCOVA using Quade's (1967) method (open triangle).
11. Post-hoc randomized block design with pretest scores as blocks (solid octagon).

For purposes of the post-hoc randomized block design, it was assumed that subjects were ranked on the basis of their pretest scores, and then blocked accordingly using three blocks. All subjects were randomly assigned to treatments independent of their pretest scores. The critical value used in all simulations was 0.05. One thousand simulations were done for each combination of correlation and effect size.

Figure 8.1 shows the results of the first simulation where equal samples sizes were in each group ($n = 10$ per group). When the effect size is 0, there should be no difference among the group means, and the percent of simulations that reject the null hypothesis should equal the critical value or α. In this case alpha was chosen to be 0.05 and each test should reject the null hypothesis 5% of the time; this is called the Type I error rate. All of the tests had Type I error rates near the nominal value of 5%. When there was no correlation between pretest and posttest, the ANOVA methods tended to have slightly liberal type I error rates near around 7%. There was a slight tendency for the error rate to increase as the correlation increased, but even then the error rate was still about 5%. None of the tests had what could be considered a high Type I error rate.

As the effect size increased, the percent of simulations which rejected the null hypothesis increased, which would be expected as the maximum difference between the group means increased. Assuming a correlation between pretest and posttest of 0.6 and an effect size difference between groups of two, the average power was about 70% for all the methods, except

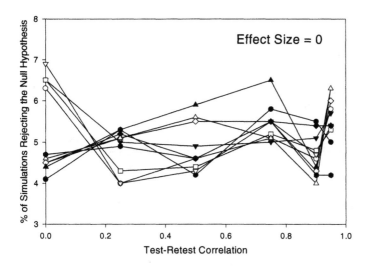

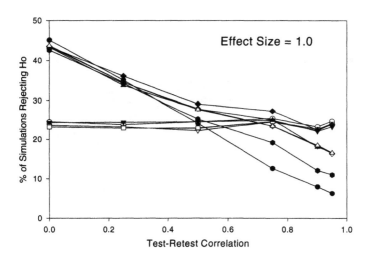

Figure 8.1a: Percent of simulations which rejected the null hypothesis (H_o: μ_1 = μ_2 = μ_3) when pretest and posttest scores were normally distributed with equal variance and equal sample sizes per group.

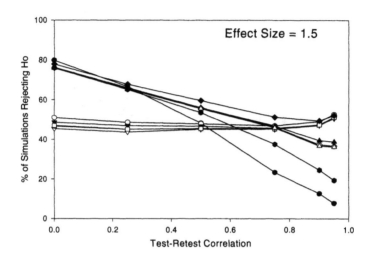

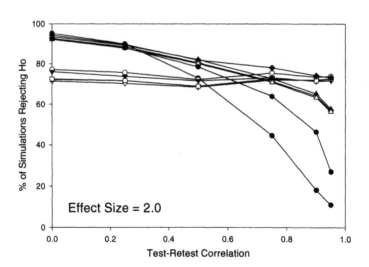

Figure 8.1b: Percent of simulations which rejected the null hypothesis (H_o: μ_1 = μ_2 = μ_3) when pretest and posttest scores were normally distributed with equal variance and equal sample sizes per group.

ANOVA on posttest scores only. In examining the plots it is important to look at those tests which stood apart from the others. Three things stood out in this simulation. One was that the power of the ANOVA on posttest scores decreased significantly as the correlation between pretest and posttest scores increased. In fact, when the effect size was two, the power of the test went from about 95% at 0 correlation to 12% at 0.95 correlation. This substantiates the claim that ignoring baseline variability in the analysis leads to lower statistical power. Second, the power of post-hoc blocking decreased significantly as the correlation between pretest and posttest decreased. In this instance, when the effect size was two, the power of the method decreased from about 0.92 at 0 correlation to 0.28 at 0.95 correlation. It remains to be seen whether this result may change had an optimal post-hoc blocking method been used rather than a fixed size block method. Third, when the correlation between tests was small ($\rho < 0.5$) there were two clusters of tests, with one cluster having much greater power than the other. The cluster with the highest power included the ANCOVA methods, whereas the cluster with the lower power included the ANOVA methods. As the correlation between pretest and posttest increased to approximately 0.75 or more, the power of the ANCOVA methods became less than the ANOVA methods, but not by much. The power of the ANOVA methods generally remained constant across all correlations.

Other observations appear as one is examining the figure. First, the post-hoc randomized block design never had greater power than ANCOVA methods regardless of the effect size or correlation and at best was equal in power to ANOVA methods. The power of the post-hoc blocking method dropped significantly compared to the ANCOVA methods when the correlation between pretest and posttest was greater then 0.75. Second, when the correlation between pretest and posttest was small to negligible ($\rho < 0.25$), ANOVA on posttest scores was just as powerful as the ANCOVA methods. What was also interesting with ANOVA on posttest scores was that as the correlation and maximal differences between group means increased, the power of the ANOVA on posttest scores dropped like a rock falling off a cliff.

In summary, when the correlation between pretest and posttest was low, near 0, there was no reason not to use ANOVA on posttest scores only. It was equally as powerful as the other methods and had the same Type I error rate. ANOVA on posttest scores only also has the advantage of ease of interpretation – when there is a difference between posttest scores there is no use in confounding the results by inclusion of pretest data in the analysis. However, when the correlation between pretest and posttest was marginal to large, ANOVA on posttest scores alone cannot be recommended as the drop in power decreases rather steeply as the correlation between pretest and posttest increases. When the relationship between pretest and posttest was linear and the distribution of the pretest and posttest scores were normally distributed, the best test appeared to be traditional ANCOVA using pretest scores as the covariate. The ANCOVA methods generally had greater power than the

ANOVA and post-hoc blocking methods, and it had the same Type I error rate as the other tests. Parametric ANCOVA with pretest scores as the covariate tended to have slightly higher power overall compared to the other ANCOVA methods. It should be stressed, however, that this conclusion is valid when all the assumptions of the ANCOVA are valid. As will be seen, this conclusion will change slightly when the assumptions are not met.

A modification of this simulation was done assuming the sample sizes in each group were unequal. In the first modification it was assumed that 5 subjects were assigned to the control group, 10 to the middle treatment group, and 15 to the treatment group with the largest treatment effect. The results of this simulation are shown in Figure 8.2. In the second modification it was assumed that 15 subjects were assigned to the control group, 10 to the middle treatment group, and 5 to the treatment group with the largest treatment effect. The second modification is simply the mirror image of the first modification with this simulation having more subjects in the control group as opposed to the treatment group. The results of the second simulation were identical to the results shown in Figure 8.2 and will not be presented. Unequal sample sizes did not affect the Type I error rate, nor did it affect the overall pattern of power curves. What was affected was the overall power rate; there was a shift in the overall power of every test downward. In both cases, unequal sample sizes decreased the absolute power of all the tests by about 20%. These simulations indicate there is no reason why unequal sample sizes should affect the choice of statistical tests because the rank order of power for any given correlation remains the same compared to equal sample sizes, but it is important to realize that unequal sample sizes affect the ability to detect treatment differences.

Monte Carlo Simulation When Systematic Bias Affects the Pretest and Posttest Equally

Two simulations were undertaken to examine the power of the tests when systematic bias affects both the pretest and posttest equally. These simulations were the same as the first simulation (n = 10 per group) with the exception that a constant bias of ± 25, e.g., $\pm 25\%$ added to the true scores, to both the pretest and posttest. Figure 8.3 presents the results of the simulation when +25 was added to each subject's true pretest and posttest scores. The results when -25 was added to the data were identical to the results when +25 was added to the data. These results will not be presented. Systematic bias had no affect on the Type I error rate, shape of the power curve, or overall power of the tests compared to Figure 8.1. This is probably because systematic bias only shifts the observed scores to a different mean value, μ_i^*, where the * denotes that it is

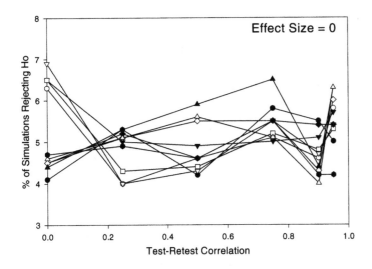

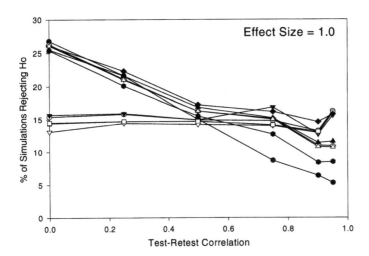

Figure 8.2a: Percent of simulations which rejected the null hypothesis when pretest and posttest scores were normally distributed with equal variance but unequal sample sizes per group. Five, 10, and 15 subjects were randomized to the control, group 2, and group 3, respectively. Groups 2 and 3 received active treatment interventions.

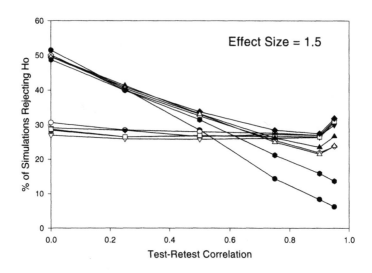

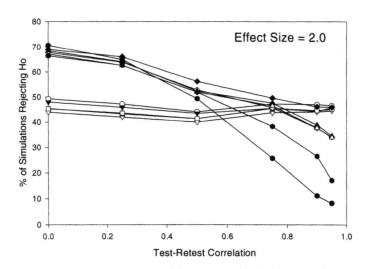

Figure 8.2b: Percent of simulations which rejected the null hypothesis when pretest and posttest scores were normally distributed with equal variance but unequal sample sizes per group. Five, 10, and 15 subjects were randomized to the control, group 2, and group 3, respectively. Groups 2 and 3 received active treatment interventions.

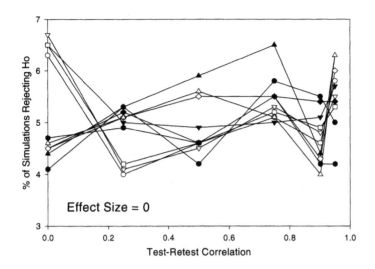

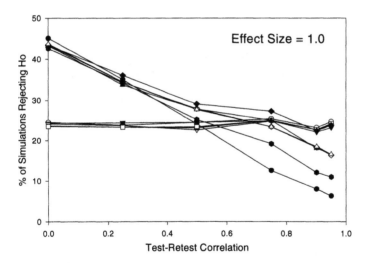

Figure 8.3a: Percent of simulations which rejected the null hypothesis when pretest and posttest scores were normally distributed with equal variance and equal sample sizes per group. However, both the pretest and posttest had a constant degree of bias of +25% in the measurement.

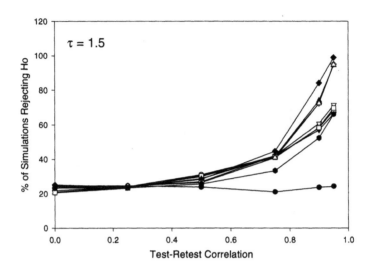

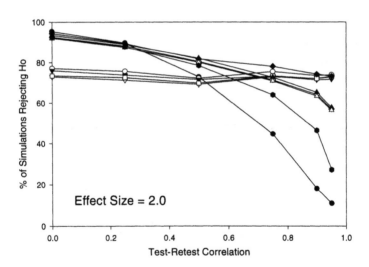

Figure 8.3b: Percent of simulations which rejected the null hypothesis when pretest and posttest scores were normally distributed with equal variance and equal sample sizes per group. However, both the pretest and posttest had a constant degree of bias of +25% in the measurement.

a biased value of the true subject value. The bias introduced in the simulation does not affect the variance of the data. Thus when systematic bias affects both the pretest and posttest scores equally one can still choose a test based on the power curve of when no systematic bias is present. What is extremely important, however, is that the degree of bias affects the pretest and posttest equally and affects all subjects equally. If either of these assumptions is violated, the power curves will change significantly, probably for the worse.

Monte Carlo Simulation When the Variance of the Posttest Scores Does Not Equal the Variance of the Pretest Scores

One of the assumptions made up until now was that the variance of the pretest was equal to the variance of the posttest. This is a reasonable assumption because if the same valid and reliable measuring device is used to measure the pretest and posttest scores within a subject, then it is reasonable to expect that the variance of the measurements would be similar. In practicality, this is not often the case. Frequently experiments will be done using a small to medium sample size and for whatever reason there will be a difference in the variance of pretest and posttest measurements. Thus the question arises as to what affect this differential variance may have on the power of the statistical test.

A Monte Carlo simulation was done to answer this question. In this simulation three groups of ten subjects each were simulated. Pretest measurements were generated having a mean of 100 and a variance of 10. Posttest scores were generated to have a mean of 100 and a variance that was a multiplier of the baseline variance. To that a constant treatment effect was added. Because the variance of the posttest was different than the variance of the pretest, the equation used to generate τ, Eq. (1.5), is not entirely valid because in that equation σ represents the common standard deviation across groups. In this simulation σ was not the same across groups. Therefore, τ was simply varied from 0 to 21. The first simulation used equal sample sizes and a posttest variance of 400, four times the pretest variance. Figure 8.4 shows the results. The Type I error among tests was similar, approximately equal to α in all cases, and was not greatly influenced by the test-retest correlation. As τ increased so did the power of the statistical tests, albeit in a curvilinear manner. For any given effect size greater than 0, the power of the tests remained relatively constant as the correlation increased up to about 0.5. Thereafter, the power of the tests increased dramatically. As an example, when the treatment effect was 7, the power of the tests was about 10-20% when the correlation was less then 0.3, but was greater than 60% when the correlation was greater than 0.9. Surprisingly, ANOVA on posttest scores alone was unaffected by the test-retest correlation. There was little difference in the power among the other statistical tests. If the variance of the pretest and posttest was equal to 100, a treatment effect of 7, 14, and 21 would be equal to an effect size of 0.7, 1.4, and 2.1, respectively. These are the conditions used in the Figure 8.1. Looking at Figure 8.1 and using interpolation, one can get a

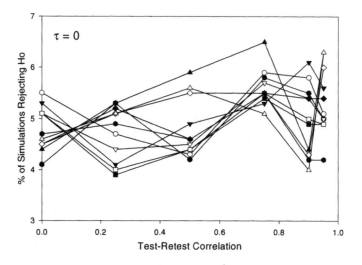

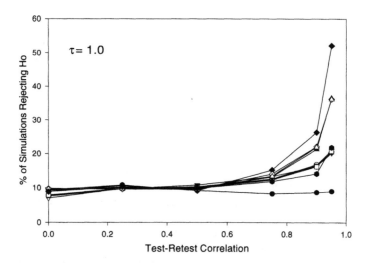

Figure 8.4a: Percent of simulations which rejected the null hypothesis when pretest and posttest samples were normally distributed but had unequal variances. The pretest mean was 100 with a variance of 100, while the variance of the posttest was 400. Ten subjects were randomized to the control group, group 2, and group 3, respectively. Groups 2 and 3 received active treatment interventions.

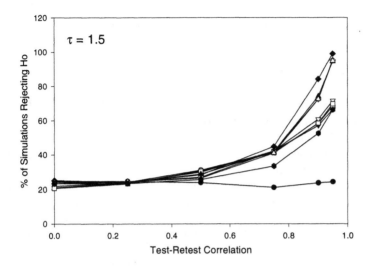

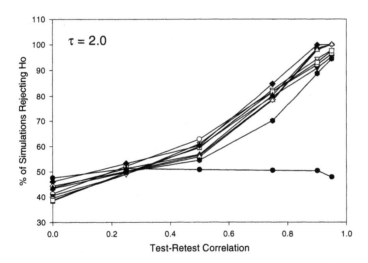

Figure 8.4b: Percent of simulations which rejected the null hypothesis when pretest and posttest samples were normally distributed but had unequal variances. The pretest mean was 100 with a variance of 100, while the variance of the posttest was 400. Ten subjects were randomized to the control group, group 2, and group 3, respectively. Groups 2 and 3 received active treatment interventions.

feel for how much power was lost when the variances of the pretest and posttest were unequal. There did not appear to be much loss in power when the effect size was small (comparing $\tau = 7$ to effect size = 1), but there was a huge drop in power when the effect size was larger than one. The power can drop as much as 50% depending on the test-retest correlation and degree of separation between pretest and posttest. These results did not change when the sample sizes between groups were different.

Monte Carlo Simulation When Subjects are Grouped *A Priori* Based on Pretest Scores (Randomized Block Designs)

All simulations that have been done up to now were based on the assumption that subjects were randomized to treatment groups without regards to pretest scores. Sometimes it is possible to analyze the pretest data before assignment of subjects to treatment groups, in which case it may be desirable to control for baseline scores through some type of blocked experimental design. We have seen two such designs in this book, the standard randomized block design (RBD) and the alternate ranks design (ARD) to be used with ANCOVA. In this simulation the power and Type I error rate of these two experimental designs were compared given that treatment assignment was based on pretest scores. Three groups of subjects were simulated with 10 subjects per group. Each subject's baseline pretest-posttest score was generated as above. In the randomized block design, subjects were sorted by ascending pretest score and grouped into blocks of 2 (open circle), 3 (solid upside down triangle), 5 (open upside down triangle), 6 (solid square), 10 (open square), and 15 (solid diamond). Within each block, subjects were randomly assigned to treatments. In the alternate ranks design, two types of assignment indicators were used. Subjects were ranked on their pretest scores and then assigned to treatments as either {1, 2, 3, 3, 2, 1, 1, 2, 3, 3, 2, 1,...etc.} or {3, 2, 1, 1, 2, 3, 3, 2, 1,...etc.}. Subjects in the ARD assigned to the first pattern of treatment assignments are referred to as Design 1 (open diamond), whereas subjects assigned to the second pattern of treatment assignments are referred to as Design 2 (solid triangle). Once subjects were assigned to treatments, the data were analyzed using parametric ANCOVA using the observed pretest score as the covariate (solid circle). The correlation of the pretest-posttest measurements was varied from 0.01 to 0.95 and the effect size was varied from 0 to 2. All assumptions of the statistical tests were met prior to analysis. The results of the simulation are shown in Figure 8.5.

There was no difference in Type I error rate among the tests and each test was near its nominal value of 5%. As expected, the power of the tests increased as the effect size increased. Within any given effect size, the first thing that was striking is that the power of the randomized block design changed as a function of the number of blocks. This would be acceptable if some type of asymptote was reached as the number of blocks increased or decreased, for then we may be able to say something like "the more blocks the

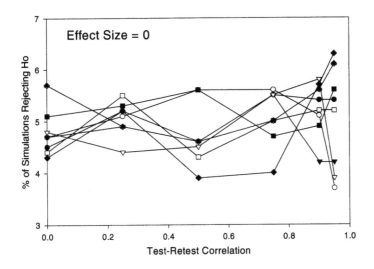

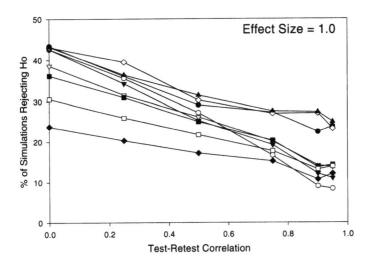

Figure 8.5a: Percent of simulations which rejected the null hypothesis when pretest and posttest scores were normally distributed and had equal sample sizes per group. Subjects were stratified on the basis of their pretest scores and randomized to blocks of different sizes (randomized block design) or randomized using the alternating block design. Ten subjects were randomized to the control group, group 2, and group 3, respectively. Groups 2 and 3 received active treatment interventions.

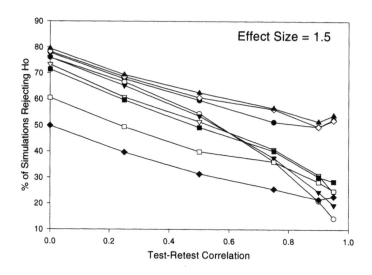

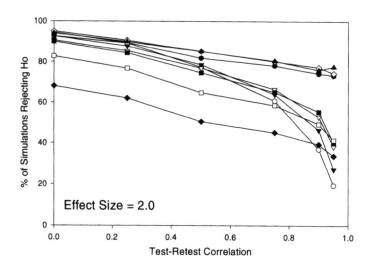

Figure 8.5b: Percent of simulations which rejected the null hypothesis when pretest and posttest scores were normally distributed and had equal sample sizes per group. Subjects were stratified on the basis of their pretest scores and randomized to blocks of different sizes (randomized block design) or randomized using the alternating block design. Ten subjects were randomized to the control group, group 2, and group 3, respectively. Groups 2 and 3 received active treatment interventions.

better," but this does not appear to be the case. The power of the randomized block design appeared to follow an inverted U-shape function with a wide range of power depending on the number of blocks chosen and the correlation between pretest and posttest. Second, as the correlation between pretest and posttest increased, the power of both the ARD and RBD decreased. For example, when the effect size was 1.5 and the correlation increased from 0 to 0.95, the power of the ARD-Design 2 decreased from 80% to 54%. It was also apparent that ARD had greater overall power than RBD, sometimes astonishingly so [compare the power of the ARD-Design 2 to the RBD(15) design when the effect size was 2.0]. There was very little difference in the power of the ARD, such that it is difficult to recommend one type of randomization sequence over the other.

The results of this simulation indicate that randomization into blocks based on pretest scores is risky. Failure to choose the "optimal" number of blocks may result in significant, decrease in statistical power. If a blocking technique must be chosen, better power is obtained with the ARD. In fact, comparing these results to Figure 8.1, the power of the ARD was slightly better than all the other methods, suggesting, that it would be prudent to use ARD whenever possible. On the down-side, the disadvantage of using the ARD as a blocking technique is that some non-statisticians may be unwilling to accept a study's results because stratifying based on alternate ranks may give the layman the impression that the investigator is attempting to bias his experiment in their favor. This is not meant to discourage the use of ARD because the use of stratification in clinical research is becoming increasingly more common.

Monte Carlo Simulation When the Marginal Distribution of the Pretest and Posttest Scores was Non-Normal

One assumption of both ANOVA and ANCOVA is that the residuals from the fitted linear model are normally distributed. This simulation was designed to determine the Type I error rate and power of each of the tests when the distribution of the residuals was not normally distributed. In practical terms, when the marginal distribution of the pretest and posttest scores are not normally distributed, the distribution of the residuals cannot be normally distributed if the sample size is large. In this simulation, 3 groups of 50 subjects each were simulated. Pretest and posttest measurements were collected on each group and were generated with a correlation of 0.6. The data were then transformed such that the marginal distribution of the pretest and posttest scores were not normally distributed. If the transformed data still did not have a correlation of 0.6 ± 0.06 then the data set was discarded. This process was iterated until a non-normally distributed data set met the correlation criterion. The reader is referred to Mooney (1997) for details. Once the marginal data set was generated, a constant treatment effect was added to each subject to simulate the effect of a treatment intervention between measurements of the pretest and posttest. The value of the treatment effect added to each subject's posttest score could not be computed using the effect size equation used previously (Cohen, 1988) because that equation is predicated on a normal, or at least symmetric distribution. Therefore, a constant effect size increasing from 0 was considered the maximal treatment effect an individual could receive. One group of individuals received the maximum treatment intervention, another received one-half the maximum treatment intervention, and the last was a placebo group which had no treatment effect added to their scores. The data were analyzed using the same statistical analyses as in the first simulation, where the assumptions of the test were met.

The non-normal distributions studied were the truncated normal distribution, log-normal, and a distribution of mixed normal distributions with 60% of the distribution in one population and 40% of the distribution in the other population. The Type I error rate and power when the residuals were normally distributed was also calculated as an internal control. These marginal distributions were chosen because each of them can be seen in typical research settings. Log-normal distributions arise when the boundary of the lower limit of the measuring device is 0, the upper limit is infinity, and the data are skewed. The truncated normal distribution arises when there is an upper and/or lower ceiling for the measuring device. For this simulation an upper and lower bound was assumed. Mixture distributions arise when in fact there are two distinct populations underlying the observed distribution. For example, suppose men and women have different means on a certain test, but this difference is not taken into account during the analysis and the two groups are pooled as a single homogenous group. This may lead to a bimodal distribution. A classic example of this might be height or weight in the general population. For this simulation, the mean of the two distributions was separated by 3 units. The general shape of each of these distributions, as evidenced by their probability distribution function, is shown in Figure 8.6.

The results of the simulation are shown in Figure 8.7. When the marginal distributions were normal, the data agreed with the results of the first simulation. Inclusion of the normal distribution as an internal control provides more validity to the results with the other distributions. It was important that the correlation of each simulation be the same within and between marginal distributions because as seen in the first simulation, when the correlation increases, the power of the test changes. Each group of simulations had similar correlations, indicating that it was valid to compare the results of each treatment effect to other treatment effects within any specific marginal distribution. Clearly, with all of the non-normal distributions, analysis of post-test scores had significantly lower power than any of the other methods. Also, the randomized block design had significantly lower power than the ANCOVA or ANOVA methods. All the methods had Type I error rates that were near

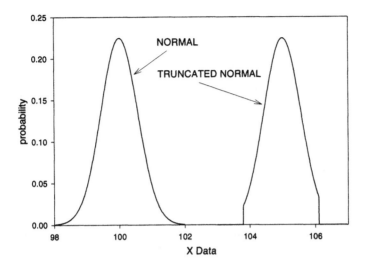

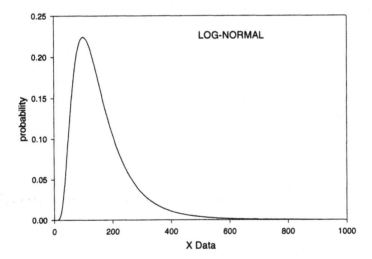

Figure 8.6a: Probability density functions for the normal, truncated, and log-normal distributions.

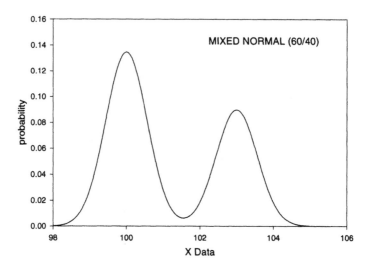

Figure 8.6b: Probability density functions for the mixture distribution.

their nominal value, regardless of the underlying distribution. All the non-normal distributions studied had power curves that were similar in shape to the normal distribution, but with power being slightly less to greatly less than the power seen with normal distribution. Like the previous simulations, among all the distributions studied, ANCOVA methods were slightly more powerful than ANOVA methods.

When the distribution was decidedly bimodal (mixed normals), the power of the ANCOVA and ANOVA methods was significantly greater than the power of the RBD method. In general, the ANCOVA methods had greater power than the ANOVA methods. The most powerful test was ANCOVA on ranked normal scores and least powerful test was analysis of posttest only scores. The Type I error rate of all the methods was near their nominal values, but was a little low overall.

When the distribution of the pretest and posttest scores was truncated normal, the power curve was almost identical to the power curve for the normal distribution. The most powerful tests were the ANCOVA methods, while the least powerful was analysis of posttest scores only. ANOVA methods were similar in power to the RBD method. Overall, parametric ANCOVA using pretest scores as the covariate was the most powerful test used. Nonparametric ANCOVA methods, while having less power than

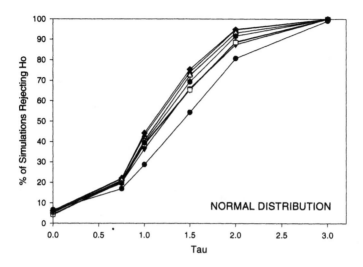

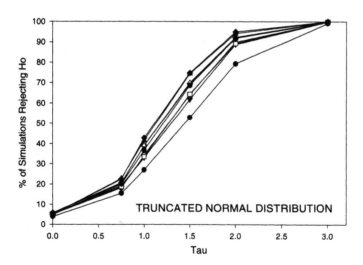

Figure 8.7a: Percent of simulations which rejected the null hypothesis when pretest and posttest scores were normally and truncated normal distributed. Fifty subjects were randomized to the control group, group 2, and group 3, respectively. Groups 2 and 3 received active treatment interventions. The correlation between pretest and posttest scores was set at $0.60 \pm 10\%$.

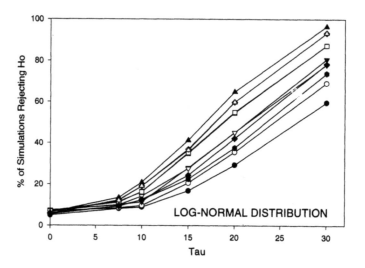

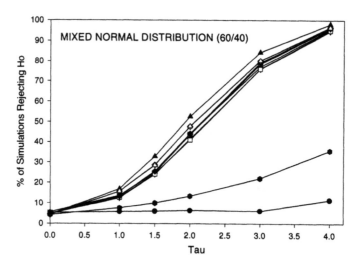

Figure 8.7b: Percent of simulations which rejected the null hypothesis when pretest and posttest scores were log-normally distributed or mixture of normal distributions. Fifty subjects were randomized to the control group, group 2, and group 3, respectively.   Groups 2 and 3 received active treatment interventions. The correlation between pretest and posttest scores was set at $0.60 \pm 10\%$.

parametric ANCOVA, had greater power than all the ANOVA methods. The Type I error rate was slightly greater than the nominal value for all tests, but only marginally so.

When the distribution of the marginals was log-normal, the most powerful tests were the ANCOVA methods, while the least powerful was analysis of posttest scores only. ANOVA methods had slightly less power than ANCOVA methods and had much greater power than the RBD method. Overall, ANCOVA using ranked normal scores as the covariate was the most powerful test studied. All the nonparametric ANCOVA methods had more power than parametric ANCOVA. The Type I error rate was higher than the nominal value and exceedingly high for the ANOVA methods.

These results support the tenet that departures from normality decrease the power of the statistical test relative to the normal distribution, sometimes drastically so. But within any particular distribution, a few conclusions can be drawn. First, analysis of posttest only scores was a complete failure for all but the truncated normal distribution. Second, RBD methods tended to have significantly less power when the distributions were quite non-normal, e.g., log-normal and mixture distribution. Third, ANCOVA methods had equal to or greater power than ANOVA methods under all conditions. Fourth, within the ANCOVA methods, ANCOVA on ranked normal scores had the greatest power overall, possibly because the methods transform the residuals back to a normal distribution. Thus ANCOVA on ranked normal scores is highly recommended for analysis of non-normal data.

Summary

- Each of the statistical tests presented in this book have slightly different interpretations.
- In presenting the results of an analysis one may wish to do an analysis using a single statistical method but present the data in an easy to understand manner.
- In general, ANCOVA methods have greater power than ANOVA and post-hoc blocking methods.
- Among the ANCOVA methods, it is best to use a nonparametric method for most analyses. These methods significantly improve the power of ANCOVA when the assumptions of the test are violated and they have only slightly lower power when the assumptions of the test are met.
- If subjects are randomized to treatment groups based on pretest scores, then the ARD should be used over the RBD.

CHAPTER 9

RANDOMIZATION TESTS

With the fall in the price of personal computers along with the concurrent increase in their processing speed, randomization tests have started to gain in their popularity among statisticians. The value of randomization tests lies in their wide applicability and their application to problems for which there is no parametric solution. Specific types of randomization tests are given names such as the bootstrap or jackknife, but some are simply called permutation tests. Surprisingly scientists in other fields have failed to grasp and make use of them. But just what are permutation or randomization tests? These tests are the generic names applied to computer-intensive methods that generate either the sampling distribution, the standard error, or p-value of a test statistic.

Good (1994) and Noreen (1989) provide two good introductory texts on permutation tests and resampling-based methods in general. Good (1994) gives a very simple outline for how to perform a permutation test

1. Analyze the problem.
2. Choose a test statistic.
3. Compute the test statistic for the observed data.
4. Rearrange (permute) the observations and recompute the test statistic for the rearranged data. Repeat until you obtain all possible permutations.
5. Accept or reject the null hypothesis using the permutation distribution as a guide.

These steps provide the basic outline for how to do permutation tests. The remainder of this section will be an outline for how to use randomization methods with some of the statistical tests that have been presented in previous chapters.

Permutation Tests and Randomization Tests

Consider a very simple example, the case where two independent groups (not paired data) are sampled and the researcher wishes to determine if the group means are equal. One method to answer this question is to compute the group means, $\overline{X}_1$ and $\overline{X}_2$, the pooled standard deviation, S_p^2, and then compute a t-test

$$t = \frac{\left|\overline{X}_1 - \overline{X}_2\right|}{\sqrt{\dfrac{s_p^2}{n} + \dfrac{s_p^2}{m}}} \tag{9.1}$$

where n and m are the sample sizes for groups 1 and 2, respectively. If t is greater than the critical value, we reject the null hypothesis of equality of group means at some level α. However, for the p-value of a test statistic to be valid, certain assumptions must be made. One assumption is that the groups have equal variance. If this assumption is violated, a correction may be made to the degrees of freedom and the corresponding critical value. Another assumption is that the groups are normally distributed. Fortunately, the t-test is quite robust to violations of this assumption. When the test statistic requires a certain underlying distribution to obtain a valid p-value then we say that that test is a parametric statistical test.

One advantage of permutation tests is that they are not parametric tests and make few to no assumptions regarding the distribution of the test statistic. This is not to say, however, that permutation tests are nonparametric tests, such as those group of test statistics which convert data to ranks before analysis and do not assume an underlying distribution of the data. In almost all cases when the parametric test applies to a problem, permutation tests have equal statistical power as the parametric test.

Permutation tests compute the sampling distribution of the test statistic directly from the observed data by exploiting the null hypothesis. Although Eq. (9.1) is referred to as the test statistic, in actuality, the test statistic is really the numerator of Eq. (9.1), the difference in the means. The denominator of Eq. (9.1) is the standard error of the numerator, which allows us to pivot the numerator into a statistic which has a known sampling distribution, i.e., the t-distribution. Permutation tests compute the standard error of a statistic and its p-value directly from the observed data.

In the example above, the null hypothesis is that the group means are equal. Suppose we were to write down the value of each observation in each group onto a separate index card and shuffle the cards many times. We then separate the cards into two new groups of size n and m. We can now compute the group means from the new groups and the difference in the group means. For simplicity, we call the new groups the resampled groups and the difference in the group means a 'pseudovalue'. If the null hypothesis were true, the difference in the resampled group means, i.e., the pseudovalue, would be close to the numerator in Eq. (9.1) because the null hypothesis states that there is no difference in the group means. Suppose we did this shuffling and recomputing of the group means over and over many times until every possible combination of data sets has been generated. This would generate a sampling distribution of pseudovalues. If we were to sort the sampling distribution in ascending order, thereby generating the cumulative distribution function for the data, we could then determine the number of observations whose pseudovalues lie greater than or equal to the observed test statistic, i.e., the numerator in Eq. (9.1). The p-value for the test statistic then becomes the number of observations greater than or equal to the observed test-statistic divided by the number of total possible combinations. It can be seen that for permutation tests to be exact and unbiased, the observations must be exchangeable. Exchangeability of observations is allowed for most problems where the data are independent and identically distributed. For the problem of determining treatment effects in the presence of baseline variables, exchangeability of observations is always assumed.

Permutation tests can actually be broken down into two different types. In one type of test (the test described previously) every possible combination of observations is generated from the observed data and the p-value is determined from the resulting sampling distribution. This type of test is referred to as a permutation test and the resulting estimate of the p-value is exact. However sometimes the data set is so large as to make every possible combination unwieldy or even impossible. From counting theory, suppose there are k groups with n_k observations per group, the number of possible combinations is

$$_nP_{n_k} = \frac{(n_1 + n_2 + n_3 + ...n_k)!}{(n_1)!(n_2)!(n_3)!...(n_k)!} \tag{9.2}$$

where ()! refers to the factorial function. Figure 9.1 plots the number of combinations in the two- and three- group case varying the number of observations per group. As can be seen from the figure, the number of possible permutations increases to astronomical levels as the number of observations per group increases every so slightly. For two groups of subjects with 15 independent subjects per group, the number of permutations is about 1.5×10^8 (Ludbrook and Dudley, 1998). For paired data, the number becomes 32,768. When the number of possible combinations increases to numbers that would be difficult to handle on a personal computer, it is necessary to reduce the number of computations by taking a small but representative sample of all possible combinations. Such a Monte Carlo approach is called a randomization test and the resulting p-value is an estimate of the true, exact p-value. Thus randomization tests represent a generalization of permutation tests and their p-values are an approximation to the true p-value. For purposes of this book, we will focus on randomization tests since they are simpler to compute and are equally valid as permutation tests when done correctly.

The utility of randomization or permutation tests should now be apparent. Foremost, a researcher is not limited to test statistics which are converted to known sampling distributions. For example, suppose the sampling distribution was quite skewed, then a test statistic, such as the difference in the modes or difference in the medians, may be needed. Further, suppose the distribution of the groups was bimodal and for whatever reason the researcher wanted to determine if the ratio of the intra-group modes was equal between groups. Certainly, there is no parametric test statistic which could provide a answer. Randomization tests could.

The disadvantage of permutation tests is devising the algorithm to test all possible combinations. For this reason, randomization tests are often used

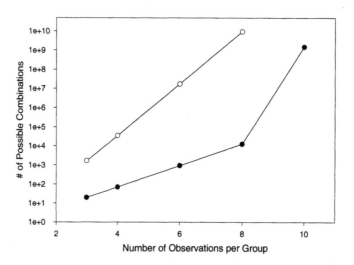

Figure 9.1: The number of possible combinations in the two- (solid circles) and three-group (open circles) case varying the number of independent observations per group.

over permutation tests. But the central question in randomization tests is how many iterations must be done to adequately simulate the sampling distribution of the test statistic. A useful tool is to plot the computed p-value at each iteration of the simulation and observe how the p-value converges to its final estimate. Often researchers use general rules of thumb and nice round numbers for how many iterations will be done. It is common to see in the literature cases where the researcher used 1000 or 10,000 iterations without any rationale for doing so. Certain computer packages, such as StatXact (Cytel Corp., Cambridge, MA), make use of permutation tests to generate p-values, thus allowing the researcher to focus on the question, not on the programming to get the answer.

In order to do more difficult analyses for which there may not be a parametric solution, it is necessary to first understand how to solve problems where a parametric solution does exist. Then by comparing the results from a known statistical test to the results from a randomization test, the reader should feel more comfortable applying randomization tests to statistics with unknown sampling distributions. To accomplish this, randomization tests for analysis of variance, analysis of covariance, and repeated measures analysis of variance will be presented with the idea being that at then end of the chapter the reader will be able to apply these methods to more complex problems.

Analysis of Variance

In an analysis of variance, the p-values reported in the ANOVA table are based on the F-distribution. The model assumes that ratio of mean squares follows an F-distribution having specific degrees of freedom. Applied to the problem at hand, estimating the treatment effect or determining the significance of a treatment in which baseline variables have been collected, we first need to determine what we are trying to answer. Recall that with randomization tests, the underlying distribution is not important provided the observations are independent and identically distributed. Thus we may make use of test statistics we defined earlier, such as difference scores or percent change, but are not limited by the inherent problems associated with these metrics. The basic idea in the k-group randomization test is to randomly shuffle the data and reassign treatments to the shuffled data. The test statistic, the treatment effect F-test, is recalculated and stored in a vector. This process is repeated many times and the proportion of pseudovalues greater than the observed test statistic is calculated. This is the p-value of the observed test statistic without regard to the F-distribution.

Figure 9.2 shows the pseudocode to the algorithm used when there are k groups. In the Appendix, there are two different sets of SAS code for the one-way analysis of variance. The first program is more general and is divided into three sections. In section one, the actual test statistic based on the observed data is generated and stored in a SAS data set called OUTF. The data is then resampled using a macro called RESAMPLE where the resampled data set is stored in a data set called NEWDATA. The test statistic is then recomputed using the macro STAT and the resultant output statistics stored in a data set called RESAMPF. The output statistics are then merged into a data set called FVALUES which contains all the F-values from the simulation. The last part of the program uses PROC IML to find the number of observations greater than or equal to the observed F-value based on the original data and uses PROC UNIVARIATE to examine the distribution of the F-values. This program can easily be extended to other situations by modifying the code contained within STAT. The second program is considerably faster and operates entirely within PROC IML. Basically, the design matrix is hard-coded within PROC IML and instead of calling PROC GLM at each iteration, the ANOVA is done using a matrix based approach. The results are stored in a vector and then the p-value is determined directly from the vector.

Consider the example in Table 9.1. In this example, three groups of unequal size had their heart rates measured before and after administration of either a placebo or two doses of a drug. Analysis of variance on the difference scores resulted in a p-value of 0.0686 [$F(2,27) = 2.96$]. The residuals from the ANOVA were not normally distributed ($p = 0.0006$) and the p-value from the ANOVA may be invalid because of violating the assumptions of the ANOVA. Analysis of variance on the rank difference scores resulted in a p-value of 0.0107 [$F(2,27) = 5.40$]. Suppose, however, that the distribution of the ratio of

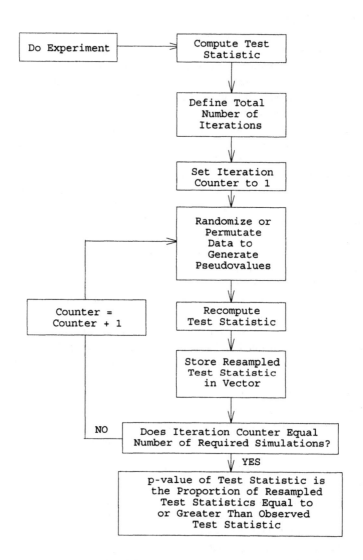

Figure 9.2: Pseudocode to the algorithm used to generate the sampling distribution for difference scores and percent change when there are k groups. The actual SAS code is given in the Appendix.

TABLE 9.1

**HEART RATE BEFORE AND AFTER ADMINISTRATION OF
PLACEBO AND TWO DOSES OF DRUG X**

	Control Group			Dose A			Dose B		
	Pretest	Posttest	Δ	Pretest	Posttest	Δ	Pretest	Posttest	Δ
	75	77	2	76	79	3	74	81	7
	68	66	-2	75	79	4	76	79	3
	82	65	-17	80	84	4	69	73	4
	76	76	0	79	73	-6	73	79	6
	73	74	1	77	81	4	77	81	4
	78	90	12	68	73	5	75	76	1
	72	68	-4	76	72	-4	71	74	3
	75	78	3	76	79	3	72	81	9
	80	79	-1	82	85	3	69	75	6
	76	75	-1	69	75	6			
				73	77	4			
Mean	76	75	-0.7	75.5	78	2.4	73	77.7	4.8
Min	68	65	-17	68	72	-6	69	73	1
Max	82	90	12	82	85	6	77	81	9

mean square's was not known, what would the p-value be? Randomization tests could be used to answer this. The data were shuffled 10,000 times and the F-value from each analysis of variance recalculated. From 10,000 iterations, 126 F-values were greater than or equal to the observed F-value of 5.40, leading to a p-value of 0.0126 (= 126/10,000).

Table 9.2 presents the output from this analysis. Figure 9.3 plots the computed p-value as a function of the iteration number. It is seen from the figure that with less than 2000 iterations, the p-value was wildly erratic varying from 0 to 0.028. Between 1000 and 3000 iterations, the calculated p-value was less variable varying between 0.012 and 0.019 and that as the number of iterations exceeded 4000, the p-value stabilized to its final value. This example demonstrates the need to have a sufficient number of iterations in the randomization process to ensure stability and validity of the computed p-value.

Analysis of Covariance

The basic linear model for analysis of covariance is

$$Y_{ij} = \mu + \tau_j + \beta(X_{ij} - \mu) + e_{ij} \qquad (9.3)$$

TABLE 9.2

STATISTICAL ANALYSIS OF DATA IN TABLE 9.1
USING RANDOMIZATION METHODS

```
          ANOVA BASED ON OBSERVED DATA (NOT RESAMPLED)
Dependent Variable: Difference Scores
                         Sum of    Mean
Source             DF    Squares   Square   F-Ratio  Prob>F
Model               2    143.66    71.83     2.96    0.0686
Error                    654.20    24.23
Corrected Total    29    797.90

Dependent Variable: Ranked Difference Scores
                         Sum of    Mean
Source             DF    Squares   Square   F-Ratio  Prob>F
Model               2    630.83    315.41    5.40    0.0107
Error              27   1578.17    58.45
Corrected Total    29   2209.00

                RESULTS OF RANDOMIZATION TEST
      Observed Test Statistic:                    5.40
      p-value Based On Resampled Distribution:    0.0126
      Number of Resampling Simulations:      10000
```

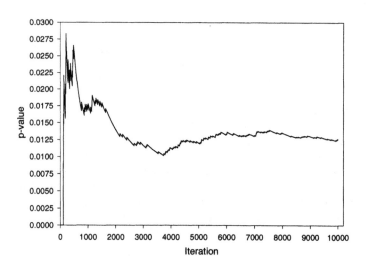

Figure 9.3: p-value as a function of iteration number. The data in Table 9.2 were analyzed using a randomization test on the difference scores. The pseudocode for the analysis is given in Figure 9.3 and the actual SAS code is given in the Appendix.

where Y_{ij} is the ith observation in the jth group, μ is the population mean, τ_j is the jth treatment effect, β is the between-groups regression coefficient, and e_{ij} is the residual associated with Y_{ij}. The equivalent randomization test for ANCOVA begins by using the recommendations of Gail, Tan, and Piantadosi (1988). They suggest fitting the analysis of covariance (ANCOVA) model to the observed data and saving the residuals. The residuals are then permutated and added back to the fitted data generating a new data set Y_{ij}^*,

$$Y_{ij}^* = \hat{Y}_{ij} + e_{ij}^* \qquad (9.4)$$

where $\hat{Y}_{ij}$ is the ith predicted value in the jth group and e_{ij}^* is the ith permuted residual in the jth group. The new permutated observations Y_{ij}^* are then submitted to analysis of covariance and the F-values recomputed. Let the recomputed F values equal F*. This process is repeated many times until a sampling distribution for F* is generated. The p-value for the observed F-value is the number of resampled F*-values greater than or equal to the observed F-value divided by the total number of resampling simulations. The SAS code to perform this analysis is given in the Appendix.

Permutation of residuals and calculation of the resulting F-value will not eliminate one very important assumption of the analysis of covariance – the between-groups regression coefficients are equal. However, if the between-group regression coefficients are equal for the observed data they will also be equal for the permutated data. Thus it is unnecessary to check for equal regression coefficients with each permutation.

As an example, consider the sexual harassment data in Table 3.4. Both pretest and posttest were transformed to ranks prior to ANCOVA. The results of resampling the residuals with 1000 permutations are given in Table 9.3. The observed F-test for treatment effect was 4.19 with a corresponding p-value of 0.0181. The resampled p-value was 0.0210, an insignificant difference. It can be seen that resampling gives a very close approximation to the p-value in the parametric case.

Resampling within Blocks or Time Periods

As we have seen in previous chapters an alternative method for dealing with baseline data is to group subjects into blocks based on their pretest scores and then analyze the posttest scores. Blocking reduces the variance of theobservations and increases the power of the statistical test. We have also seen that in a repeated measures design, multiple posttest measurements are made on the same subject. One assumption of randomization tests is that the observations are exchangeable. To simply randomize data between blocks or time periods would violate this assumption. Randomization must be done within blocks or time periods for exchangeability to still be valid. In order to do a randomization test involving blocked or time-dependent data, first the

TABLE 9.3

**ANALYSIS OF SEXUAL HARASSMENT DATA (TABLE 3.4) USING
ANCOVA RANDOMIZATION TECHNIQUES
AFTER RANK TRANSFORMATION OF THE DATA**

Results of randomization test	
Observed Test Statistic (F-value):	4.1940
p-value Based On Resampled Distribution:	0.0210
Number of Resampling Simulations:	1000

Univariate summary statistics for resampled F-values			
N	1000	Sum Wgts	1000
Mean	1.040384	Sum	1039.343
Std Dev	1.119778	Variance	1.253903
Skewness	2.245445	Kurtosis	7.275795
CV	107.6313	Std Mean	0.035428
W:Normal	0.780203	Pr<W	0.0001

observed test statistic is calculated. Second, subjects are rearranged at random within blocks or time periods, taking care that the number of observations in each treatment category remains constant. From this point on randomization proceeds as before.

Resampling with Missing Data
In the examples presented previously, there were no missing data. However, in any clinical study it is usually the rule, rather than the exception, that missing data will occur within subjects. The effects of missing data on the analysis will depend on the nature of the experimental design. Randomization tests may still be used in the case of k-group pretest-posttest designs because missing data will have no impact on the outcome other than a decrease in statistical power. Missing data is a real problem with more complex experimental designs with multiple factors. If any entire level of a factor is missing, it may not be possible to estimate any of the treatment effects. In such an instance it may be incorrect to use randomization methods. The reader is referred to Good (1994) for dealing with missing data in complex situations.

Summary
• Randomization tests are not a panacea for all problems inherent in certain tests.

- Resampling is a powerful tool available to the researcher, although it requires greater programming skill than simple statistical analyses (as can be seen by the SAS code given in the Appendix), because it can be extremely helpful in situations where a parametric solution is unavailable or there is uncertainty regarding the validity of the assumptions of the parametric test.
- The researcher should be aware that other more sophisticated techniques are available for pretest-posttest analysis if they are needed.
- Many of the topics in this book were not covered due to space limitations, such as how randomization can be applied to the problem of equality of variances.

CHAPTER 10

EQUALITY OF VARIANCE

Methods and Procedures

Up to now the problem at hand has been to determine whether the posttest means were equal among treatment groups after taking into account the dependency on pretest scores. For many of the statistical analyses presented, it was assumed that the variance of the pretest measurements was equal to the variance of the posttest measurements. Only with one of the simulations in the previous chapter did we test to see how the power of the statistical tests presented fared when the variances were unequal. Sometimes the assumption of equality of variance may be in doubt and it becomes necessary to test this assumption directly. The null and alternative hypotheses may be written as

$$H_o : \sigma_X^2 = \sigma_Y^2$$
$$H_a : \sigma_X^2 \neq \sigma_Y^2$$

or equivalently

$$H_o : \sigma_X^2 - \sigma_Y^2 = 0$$
$$H_a : \sigma_X^2 - \sigma_Y^2 \neq 0$$

where X and Y refer to the pretest and posttest measurements, respectively.

Pittman (1939) and Morgan (1939) first tackled this problem more than 50 years ago. They showed that if $E(X) = \mu_X$, $E(Y) = \mu_Y$, $Var(X) = \sigma_X^2$, $Var(Y) = \sigma_Y^2$, and the correlation between X and Y was equal to ρ, then the covariance between the difference scores,

$$D_i = Y_i - X_i \tag{10.1}$$

and sum scores,

$$S_i = Y_i + X_i \tag{10.2}$$

can be written as

$$
\begin{aligned}
Cov(D,S) &= Cov(Y-X, X+Y) \\
&= Var(X) + Cov(X,Y) - Cov(X,Y) - Var(Y) \\
&= \sigma_X^2 - \sigma_Y^2
\end{aligned}
\tag{10.3}
$$

Notice that the right side of Eq. (10.3) is equal to the form of the null hypothesis to be tested. If we can find a test statistic which tests whether the covariance of the difference and sum scores equals 0, this would be equivalent to a test of whether the variances of X and Y were equal.

Recall that the correlation between two variables is the ratio of their covariance to the product of their individual standard deviations. Then the correlation between the difference and sum scores, ρ_{sd}, is

$$\rho_{sd} = \frac{\text{cov}(D,S)}{\sigma_d \sigma_s} = \frac{\sigma_X^2 - \sigma_Y^2}{\sigma_d \sigma_s}, \qquad (10.4)$$

where σ_d and σ_s are the standard deviations of the difference and sum scores, respectively. Eq. (10.4) is also similar to the null hypothesis to be tested with the exception of the presence of a denominator term. The denominator term maps the domain of the covariance from the interval $(-\infty, \infty)$ to the interval $(-1, 1)$. Thus a test of H_o: $\sigma_X = \sigma_Y$ is equivalent to a test for 0 correlation between the sum and difference scores.

McCullogh (1987) showed that Pittman's method is not robust to departures from normality. When the tails of the marginal distributions are heavy, the Type I error rate is over-estimated. Thus McCullogh (1987) proposed that Spearman's rank correlation coefficient be used instead of the Pearson product moment correlation coefficient in testing for equality of variances. Using Monte Carlo simulation, McCullogh (1987) showed that Spearman's rank correlation coefficient was much better at controlling the Type I error rate than the Pearson product moment correlation coefficient. One advantage of both Pittman's method and McCullogh's alternative is that they can easily be calculated using any statistics package.

Bradley and Blackwood (1989) presented an interesting modification of Pittman's test. If one considers the conditional expectation of the sum scores (S) given the difference scores (D), then

$$E(D|S) = (\mu_X - \mu_Y) + \rho_{ds} \frac{\sigma_d}{\sigma_s} [S - (\mu_X + \mu_Y)]$$

$$E(D|S) = (\mu_X - \mu_Y) + \left[\frac{\sigma_X^2 - \sigma_Y^2}{\sigma_S^2} \right] [S - (\mu_X + \mu_Y)] \qquad (10.5)$$

$$E(D|S) = \beta_0 + \beta_1 S$$

where

$$\beta_0 = (\mu_X - \mu_Y) - \left[\frac{\sigma_X^2 - \sigma_Y^2}{\sigma_S^2} \right] (\mu_X + \mu_Y)$$

and

$$\beta_1 = \frac{\sigma_X^2 - \sigma_Y^2}{\sigma_S^2}.$$

$\sigma_X^2 = \sigma_Y^2$ and $\mu_X = \mu_Y$ if and only if $\beta_0 = 0$ and $\beta_1 = 0$. Therefore, a simultaneous test for equal means and variances of paired data can be calculated from the linear regression of difference scores against sum scores. The test statistic

$$F = \frac{\left(\sum_{i=1}^{n} D_i^2 - SSE \right)(n-2)}{2 \cdot SSE} \qquad (10.6)$$

where SSE is the error sum of squares obtained from the linear regression, may be used to test the null hypothesis of equal means and variances. Reject H_o if $F > F_{crit}$ where F_{crit} is the critical value obtained from the $1-\alpha$ quantile of an F-distribution with 2 and n-2 degrees of freedom.

As an example, consider the sexual harassment data presented in Table 3.4. The variance of the pretest scores was 537, whereas the variance of the posttest scores was 719. The Pearson product-moment correlation between the sum and differences was 0.1603 ($p = 0.119$) and the Spearman rank correlation was 0.1141 ($p = 0.268$). Because the p-value for the correlation coefficient was greater than 0.05, the hypothesis of equal variance was not rejected at $\alpha = 0.05$.

Sandvik and Olsson (1982) proposed an alternative robust method to test for equality of variances for paired data. If we denote the median of the pretest and posttest scores as M_X and M_Y, respectively, then we can define a new variable, Z_i, for both the pretest and posttest as

$$Z_i = |X_i - M_X| - |Y_i - M_Y|, \; i=1, 2, ...,n \qquad (10.7)$$

where $|.|$ denotes the absolute value function. The Z's will tend to be positive under the alternative hypothesis that $\sigma_X^2 > \sigma_Y^2$. Thus Sandvik and Olsson (1982) proposed using Wilcoxon's signed-rank test on the Z's to test the null hypothesis. Monte Carlo simulation showed that the Wilcoxon's alternative was equally valid as Pittman's method when the marginal distributions were normally distributed and it performed much better when the data were even moderately skewed. Grambsch (1994) used extensive Monte Carlo simulation to compare Pittman's method, Pittman's method using Spearman's rank correlation coefficient, Sandvik and Olsson's (1982) robust method, and another robust method developed by Wilcox (1989). Grambsch (1994) tested the power of the statistics when the marginal distribution was uniform, normal, double exponential, slash, beta, exponential, or chi-square and the null hypothesis was either true or false. As expected, across all statistical tests, power tended to decrease from a symmetrical to skewed distribution and from a lighter to heavier tailed distribution. The author concluded that no test was superior under all conditions and at least one method was superior under at least one assumption/distribution simulation. However, the author did suggest that if one test had to be chosen from amongst the others, then Pittman's test using the Spearman rank correlation coefficient has the most appeal.

It would probably be prudent to use the Pittman-Morgan test with Spearman's rank correlation coefficient as the test of choice for a number of reasons. First, Grambsch's (1994) research demonstrated the superiority (or at least equality) of the Pittman-Morgan test with Spearman's rank correlation coefficient under a variety of different conditions. Second, statisticians have a long history of experience with the Pittman-Morgan test and it is a widely recognized statistical test, as opposed to the test by Sandvik and Olsson (1982) which most statisticians are probably not familiar with. This is not to say that one shouldn't use a test because it is unheard of or people are unfamiliar with, but rather to keep in mind that the test is unproven and its use should be done with caution. Third, in this example the marginal distribution of the posttest was not normally distributed, suggesting that some nonparametric alternative to the traditional Pittman-Morgan test be used. For these reasons, it should be concluded that the variances were equal.

The methods just presented are useful when there are only two groups, the pretest and posttest groups. Wilcox (1989) provided a statistic to test whether the variance of q groups with pretest-posttest data have equal variance. One example of this may be in a repeated measures design with baseline measurement where it may be necessary to test whether the variances are equal at each measurement period. The null and alternative hypotheses are

$$H_o : \sigma_1^2 = \sigma_2^2 = ...\sigma_j^2, \ j=1, 2,...q$$

$$H_a : \text{at least one } \sigma_i^2 \neq \sigma_j^2, i \neq j$$

Using Sandvik and Olsson's transformation (1982), Eq. (10.7), let $R(Z_{ij})$ be the rank of Z_{ij} within the ith subject. For example, suppose each subject is measured on some variable once on each of three occasions and the scores for the first subject are 10, 30, and 20, for the first, second and third time period, respectively. Then i = 1, $R(Z_{11}) = 1$, $R(Z_{12}) = 3$, and $R(Z_{13}) = 2$. Let

$$D_i = \max\{Z_{ij}\} - \min\{Z_{ij}\}, i=1, 2, ...n \tag{10.8}$$

be the difference between the largest and smallest Z_{ij} within each subject. For the example above, $D_1 = 30-10 = 20$. Let $R(D_i)$ be the rank of D_i. Now compute the following

$$S = \sum_{j=1}^{k} \left[\sum_{i=1}^{n} \left(R(D_i) \cdot R(Z_{ij}) \right) \right]^2 \tag{10.9}$$

$$W = \frac{72 \cdot S}{k(k+1)n(n+1)(2n+1)} - \frac{9(k+1)n(n+1)}{2(2n+1)} \tag{10.10}$$

$$A = 1 - \frac{6(3n^2 + 3n - 1)}{5n(n+1)(2n+1)} \tag{10.11}$$

$$B = 3A - 2 + \frac{72(3n^4 + 6n^3 - 3n + 1)}{7n^2 (n+1)^2 (2n+1)^2} \tag{10.12}$$

$$v = \frac{(k-1)A^3}{B^2} \tag{10.13}$$

$$Q = \frac{A (W - J + 1)}{B} + v . \tag{10.14}$$

Reject H_o when $Q > Q_{crit}$ where Q_{crit} is the critical value associated with a $1-\alpha$ quantile of a chi-square distribution with v degrees of freedom rounded to the nearest integer.

As an example of the method, three groups of subjects were generated with ten subjects per set. Two of the groups were distributed as $X \sim N(10,1)$ and one group was generated as $X \sim N(10,5)$. The data are presented in Table 10.1. The following values were calculated using Eq. (10.9)-(10.14): $S = 39794$, $W = 9.07$, $A = 0.8291$, $B = 0.5566$, $v = 3.67$, $Q = 14.218$, and $p(Q, 4) = 0.0026$. Thus the null hypothesis was rejected and it was concluded that at least one of the group variances was not equal to the others. One disadvantage to using the Q method is that no data can be missing. If any data are missing that entire subject must be deleted from the analysis. Another disadvantage is that it fails to identify which time period has unequal variance compared to the others or whether two or more time periods have unequal variance compared to the rest. Nevertheless, it is still a useful test for pointing out that there is heteroscedasticity present in the data set on hand.

Summary
- Methods were presented for null hypothesis testing for equality of the pretest and posttest variance.
- Until further research is done using the different methods under varying conditions, it would probably be wise to use the Pittman-Morgan method modified using the Spearman rank correlation coefficient.

TABLE 10.1

EXAMPLE OF WILCOX'S Q METHOD FOR TESTING THE
EQUALITY OF VARIANCE AMONG k GROUPS

Subject	Original Observations			Transformed Observations			
	X_1	X_2	X_3	Z_1	Z_2	Z_3	D
1	9.19	10.44	11.81	0.36	0.46	1.83	1.47
2	9.50	10.39	10.59	0.04	0.41	0.60	0.56
3	9.19	10.05	3.45	0.35	0.07	6.53	6.46
4	10.02	9.34	10.03	0.48	0.64	0.05	0.58
5	10.42	9.35	14.40	0.87	0.63	4.42	3.79
6	9.59	9.00	9.93	0.04	0.98	0.05	0.94
7	10.66	11.11	8.18	1.12	1.13	1.80	0.69
8	9.35	9.91	8.97	0.19	0.07	1.02	0.95
9	11.05	9.09	12.61	1.51	0.89	2.63	1.74
10	8.51	10.51	4.66	1.04	0.54	5.32	4.79
mean	9.75	9.92	9.46				
variance	0.49	0.49	11.42				

Subject	Rank Transformation				$R(Z_i) * R(D_i)$		
	$R(Z_1)$	$R(Z_2)$	$R(Z_3)$	R(D)			
1	1	2	3	6	6	12	18
2	1	2	3	1	1	2	3
3	2	1	3	10	20	10	30
4	2	3	1	2	4	6	2
5	2	1	3	8	16	8	24
6	1	3	2	4	4	12	8
7	1	2	3	3	3	6	9
8	2	1	3	5	10	5	15
9	2	1	3	7	14	7	21
10	2	1	3	9	18	9	27
			sum of columns		96	77	157
		sum of columns squared			9216	5929	24649
	grand sum of columns squared					39794	

APPENDIX

SAS Code

```
/***********************************************************
*    SAS Code to Do:
*        Iteratively Reweighted ANCOVA
***********************************************************/
%MACRO IRWLS;
  proc sort;
    by resid;
  proc iml;
    use resid;
    read all;
    absresid = abs(resid);
    if nrow(resid)/2 = int(nrow(resid)/2) then;
        mad = (absresid(nrow(resid)/2) +
            absresid(nrow(resid)/2 + 1))/(2 * 0.6745);
    else;
        mad = absresid(nrow(resid)/2 + 1)/0.6745;
    u = resid/mad;
    huberwgt = j(nrow(resid), 1, 1);
    bisqr = j(nrow(resid), 1, 0);
    do j = 1 to nrow(resid);
        if abs(u(j)) > 1.345 then
            huberwgt(j) = 1.345/abs(u(j));
        if abs(u(j)) <= 4.685 then
            bisqr(j) = (1 - (u(j)/4.685)**2)**2;
    end;
    create data2 var {trt pretest posttest huberwgt bisqr};
    append var {trt pretest posttest huberwgt bisqr};
  run;
%MEND IRWLS;

%MACRO BISQRGLM;
  proc glm data=data2;
    class trt;
    model posttest = trt pretest;
    weight bisqr;
    output out=resid r=resid;
    title1 'GLM with Bisquare Function Weights';
  run;
%MEND BISQRGLM;

%MACRO HUBERGLM;
  proc glm data=data2;
    class trt;
    model posttest = trt pretest;
    weight huberwgt;
    output out=resid r=resid;
    title1 'GLM with HUBER Function Weights';
  run;
%MEND HUBERGLM;

data original;
  input trt pretest posttest;
  cards;
    0      72      74
```

179

```
        0    74    76
        0    76    84
        0    70    67
        0    71    79
        0    80    72
        0    79    69
        0    77    81
        0    82    72
        0    69    60
        1    91   112
        1    75   106
        1    74   104
        1    83    96
        1    81    19
        1    72    99
        1    75   109
        1    77    98
        1    81   106
;

/* do Huber function 2 times starting with OLS estimates */
proc glm data=original;
  class trt;
  model posttest = trt pretest;
  output out=resid r=resid;
  title1 'GLM with OLS estimates';
%IRWLS;
%HUBERGLM;
%IRWLS;
%HUBERGLM;

/* do bisquare function 2 times starting with OLS estimates*/
proc glm data=original;
  class trt;
  model posttest = trt pretest;
  output out=resid  r=resid;
  title1 'GLM using OLS estimates';
%IRWLS;
%BISQRGLM;
%IRWLS;
%BISQRGLM;



/*************************************************************
* SAS Code to do:
*       ANCOVA WHEN THE WITHIN-GROUP
*       REGRESSION COEFFICIENTS ARE UNEQUAL
*************************************************************/
data rawdata;
  input group posttest pretest;
  cards;
  1 16 26
  1 60 10
  1 82 42
  1 126 49
  1 137 55
  2 44 21
  2 67 28
  2 87 5
  2 100 12
  2 142 58
  3 17 1
```

```
    3  28  19
    3 105  41
    3 149  48
    3 160  35
;
/*    Compute Quade's Test */
proc iml;
  use rawdata;
  read all var {group pretest posttest};
  rankpre = ranktie(pretest);
  rankpost = ranktie(posttest);
  devrankx = (rankpre-sum(rankpre)/nrow(rankpre));
  devranky = (rankpost-sum(rankpost)/nrow(rankpost));
  b = inv(devrankx` * devrankx) * devrankx` * devranky;
  yhat = devrankx * b;
  residual = devranky - yhat;
  create resdata  var {group pretest posttest rankpre
    rankpost yhat residual};
  append var {group pretest posttest rankpre
    rankpost yhat residual};
run;

title 'Quade test for Nonparametric Analysis of Covariance';
proc glm data=resdata;
  class group;
  model residual=group;
run;

title 'Test for Heterogeneity of Slopes';
proc glm;
class group;
  model posttest=group pretest group*pretest;
run;

title 'Parametric Analysis of Covariance';
proc glm;
  class group;
  model posttest=group pretest;
run;

title 'Parametric Analysis of Covariance on Ranks';
proc glm;
  class group;
  model rankpost = rankpre group;
run;


/************************************************************
*
*   SAS Code to :
*      COMPUTE G FOR Ln TRANSFORMATION
*
*************************************************************/
dm 'clear log';
dm 'clear list';


data one;
        input x1 x2;
        cards;
-10 25
5 28
7 45
-8 32
```

```
0  4
-14 18
2  15
-10 14
12 -4
7  9
18 24
-11 23
-1 10
0  16
-2 18
4  2
7   8
10   16
-4 24
0  34
2  32
3  16
-6 17
-7 -1
10 11
8  14
0   25
1    19
-4   18
2   16
-4 15
2  7
5  -1
-1 3
0  -2
13 27
9  26
5  32
;

proc iml;
   use one;
   read all var {x1 x2};
   poolx = x1||x2;
   minx = abs(min(poolx)) + 0.1;
   n = nrow(x1);
   ming = 10000;
   c = minx + cusum(j(100 * minx, 1, 0.1));
   g = j(nrow(c), 1, .);
   do i = 1 to nrow(c);
        l = log(x1 + c(i)) - log(x2 + c(i));
        sy = sum(l);
        sy2 = sum(l##2);
        sy3 = sum(l##3);
        sy4 = sum(l##4);
        s = (sy2 - sy##2/n)/(n-1);
        k3 = (n#sy3 - 3#sy#sy2 + 2#sy##3/n)/((n-1)#(n-2));
        g1 = k3/s##3;
        k4 = ((n##3 + n##2)#sy4 - 4#(n##2+n)#sy3#sy -
                3#(n##2-n)#sy2##2 + 12#n#sy2#sy##2 -
                6#sy##4) / (n#(n-1)#(n-2)#(n-3));
        g2 = k4/s##4;
        g(i) = abs(g1) + abs(g2);
        if g(i) <= ming then do;
             optc = c(i);
             ming = g(i);
           end;
   end;
```

```
    print 'Optimum value of c: ' optc;
    print 'Minimum value of g: ' ming;
    l = log(x1 + optc) - log(x2 + optc);
    create logtran var {x1 x2 l};
    append var {x1 x2 l};
    create optmum var {c g};
    append var {c g};
run;

proc univariate data=logtran normal;
    var l;
run;


/**********************************************************
 *
 *   SAS Code to:
 *     PROCEDURE TO RESAMPLE WITHIN K-GROUPS (ANOVA Type)
 *
 **********************************************************/
options ls=75;
dm 'clear log';
dm 'clear list';

data data;
        input trt subject pretest posttest;
        diff = posttest - pretest;
        cards;
1 1 75 77
1 2 68 66
1 3 82 65
1 4 76 76
1 5 73 74
1 6 78 90
1 7 72 68
1 8 75 78
1 9 80 79
1 10 76 75
2 11 76 79
2 12 75 79
2 13 80 84
2 14 79 73
2 15 77 81
2 16 68 73
2 17 76 72
2 18 76 79
2 19 82 85
2 20 69 75
2 21 73 77
3 22 74 81
3 23 76 79
3 24 69 73
3 25 73 79
3 26 77 81
3 27 75 76
3 28 71 74
3 29 72 81
3 30 69 75
;

proc rank data=data out=data;
        var diff;
        ranks rdiff;
run;
        /*******************************************
```

```
    This is a macro that can be used to resample
k-groups of scores. In this instance,
ranked difference scores will be resampled.
************************************/
%MACRO RESAMPLE;
    proc iml;
    use data var _all_;
    read all;
    /**************************************************
    modify the next uncommented line by changing
        mat = rdiff to  mat = VARIABLE
    to reflect the dependent variable being resampled
    ****************************************************/
                  y = j(nrow(rdiff), 1, .);
                  call ranuni(0, y);
                  mat = rdiff||y;
                  sortmat = j(nrow(mat), ncol(mat), .);
                  sortmat[rank(mat[, 2]), ] = mat[];
                  y = sortmat[1:nrow(sortmat), 1];
    create newdata var {trt y};
    append var {trt y};
%MEND RESAMPLE;

/****************************************************
    MACRO STAT reflects the  type of analysis
    In this case, an analysis of variance is being
    done
    ****************************************************/
%MACRO STAT(DV);
    proc glm data=newdata outstat=outstat noprint;
        class trt;
        model &DV = trt;
        run;
%MEND STAT;

%MACRO SIM;
  data fvalues; f =.; id = .;
  %do i = 1 %to 10000;
            dm 'clear log';
            %let id = &i;
            %resample;
            %stat(Y);
            data f&i;
                set outstat;
                id = &i;
                resampf = f;
                    if _type_ = 'SS3' and _source_ = 'TRT'
                        then output;
                drop _type_ _source_ df prob _name_ ss;
            data fvalues;
                merge fvalues f&i;
                by id;
                if id ^= . then output;
                drop f;
            proc datasets;
                delete f&i;
    %end;
  %MEND SIM;

/* compute the observed F-statistic */
proc glm data=data outstat=obsf;
        class trt;
        model rdiff = trt;
run;
```

```
        data obsf;
            set obsf;
            if _type_ = 'SS3' and _source_ = 'TRT' then output;
            drop _type_ _source_ df prob _name_ ss;
        %sim;
        proc iml;
            use fvalues var _all_;
            read all;
            use obsf var _all_;
            read all;
            flt = choose(resampf < f, resampf, .);
            fgt = setdif(resampf, flt)';
            n = nrow(resampf);
            p = nrow(fgt)/n;
            reset center noname;
            mattrib f format=8.4;
            mattrib p format=8.4;
            mattrib n format=8.0;
            print 'Results of Randomization Test';
            print 'Observed Test Statistic (F-value):' f;
            print 'p-value Based On Resampled Distribution:'p;
            print 'Number of Resampling Simulations:' n;
        run;


/************************************************************
*
*   OPTIMIZED SAS Code to:
*       PROCEDURE TO RESAMPLE WITHIN K-GROUPS (ANOVA Type)
*       ONE-WAY ANALYSIS OF VARIANCE
*       DESIGN MATRIX HARD CODED WITHIN PROC IML
*************************************************************/
data data;
        input trt subject pretest posttest;
        diff = posttest - pretest;
        pchange = (posttest - pretest)/pretest * 100;
        cards;
1 1 75 77
1 2 68 66
1 3 82 65
1 4 76 76
1 5 73 74
1 6 78 90
1 7 72 68
1 8 75 78
1 9 80 79
1 10 76 75
2 11 76 79
2 12 75 79
2 13 80 84
2 14 79 73
2 15 77 81
2 16 68 73
2 17 76 72
2 18 76 79
2 19 82 85
2 20 69 75
2 21 73 77
3 22 74 81
3 23 76 79
3 24 69 73
3 25 73 79
3 26 77 81
3 27 75 76
3 28 71 74
```

```
3 29 72 81
3 30 69 75
;
proc glm data=data outstat=obs;
    title1 'ANOVA on Ranked Difference Scores';
    class trt;
    model diff = trt;
    means trt / regwq;
    output out=res residual=res;
run;
data obs(keep=obsf);
    set obs;
    obsf = f;
    if _source_ = 'TRT' and _type_ = 'SS3' then output;

title1 'p-value obtained by Randomization';
title2 'Summary Statistic is the F-test from Ranked Difference
Scores';
proc iml;
    use data;
    read all var _all_;
    use obs;
    read all var {obsf};

    /* begin randomization */
    nosim = 10000;
    x = {1 1 0, 1 1 0, 1 1 0, 1 1 0, 1 1 0, 1 1 0, 1 1 0, 1 1 0,
         1 1 0, 1 1 0, 1 0 1, 1 0 1, 1 0 1, 1 0 1, 1 0 1, 1 0 1,
         1 0 1, 1 0 1, 1 0 1, 1 0 1, 1 0 1, 1 0 0, 1 0 0, 1 0 0,
         1 0 0, 1 0 0, 1 0 0, 1 0 0, 1 0 0, 1 0 0};
    f = j(nosim, 1, .);
    do simctr = 1 to nosim;
            y = j(nrow(rdiff), 1, .);
            call ranuni(0, y);
            mat = diff||y;
            sortmat = j(nrow(mat), ncol(mat), .);
            sortmat[rank(mat[, 2]), ] = mat[];
            y = sortmat[1:nrow(sortmat), 1];
            b = inv(x` * x) * x` * y;
            sse = y` * y - b` * x` * y;
            sstrt =  b`*x`*y - sum(y)##2/nrow(y);
            dfe = nrow(y) - 1 - 2;
            mstrt = sstrt/2;
            mse = sse/dfe;
            if mse < 0 then f[simctr] = 0;
                    else f[simctr] = mstrt/mse;
    end;
    simctr = cusum(j(nosim, 1, 1));
    pvalue = choose(f >= obsf, 1, 0);
    pvalue = cusum(pvalue)/simctr;
    obsf = j(nosim, 1, obsf);
    create fvalues var{f pvalue simctr obsf};
    append var {f pvalue simctr obsf};

data pvalue;
    set fvalues;
    if f > obsf then output;
    proc means data=pvalue n;
        title3 'Number of Simulations Greater than Observed F';
        var f;
    run;
```

```
/************************************************************
*    SAS Code to:
*    RESAMPLE WITHIN K-GROUPS WITH T-DIFFERENT TIME POINT
*       (Repeated Measures Type of Analysis)
*************************************************************/
     dm 'clear log';
     dm 'clear output';

     proc format;
         value sequence 1='RT'  2='TR';
     run;

     data one;
         infile rmdata;
         input sequence subject day period trt cs baseline;
         diff = cs - baseline;

     %MACRO RESAMPLE;
         proc sort data=one; by day;
         proc iml;
           seed1 = time();
           use one var _all_;
           read all;
           /************************************************
               modify the next uncommented line by changing
                   times = day to  times = VARIABLE
               to reflect the variable that is being repeated
           ************************************************/
         times = day;
           /************************************************
               modify the next uncommented line by changing
                   newy = diff to  newy = VARIABLE
               to reflect the dependent variable
           ************************************************/
         newy = diff;
           timevar = unique(times)';
           do i = 1 to nrow(timevar);
                 ti = choose(times ^= timevar(i) , times, .);
                 y = .;
                 do k = 1 to nrow(ti);
                       if ti(k) ^= . then y = y//newy(k);
                 end;
                 y = y(2:nrow(y));
                 seed1 = time();
                 u = rank(ranuni(J(nrow(y), 1, seed1)));
                 y = y(|u,|);
                 /**********************************
                     modify the next uncommented line by changing
                         if i = 1 then diff = y;
                             else diff = diff//y;
               to
                         if i = 1 then diff = VARIABLE;
                             else diff = diff//VARIABLE;
                     to reflect the dependent variable
                 **************************************/
                 if i = 1 then diff = y;
                     else diff = diff//y;
     create newdata var {sequence subject trt period
         day diff};
      append var {sequence subject trt period day diff};
     run;
     %MEND RESAMPLE;
```

```
/*********************************************************
      MACRO STAT reflects the   type of analysis
        In this case, an analysis of variance is being
      done
      *******************************************************/
%MACRO STAT;
    proc glm data=newdata outstat=outstat noprint;
            class sequence trt subject period trt day;
            model diff = sequence subject(sequence)
            period trt day;
    run;
%MEND STAT;

%MACRO SIM;
    data fvalues; f =.; id = .;
    /*******************************************
       change the counter (i) to the number
         of desired simulations
       *****************************************/
       %do i = 1 %to 1200;
            %let id = &i;
            %resample;
            %stat;
            data ftrt&i;
                set outstat;
                id = &i;
                if _type_ = 'SS3' and _source_ = 'TRT'
                then ftrtrs = f;
                if _type_ = 'SS3' and _source_ = 'TRT'
                then output;
                drop _type_ _source_ df prob _name_ ss f;
            data fday&i;
                set outstat;
                id = &i;
                if _type_ = 'SS3' and _source_ = 'DAY'
                then fdayrs = f;
                if _type_ = 'SS3' and _source_ = 'DAY'
                then output;
                drop _type_ _source_ df prob _name_ ss f;
            proc sort data=ftrt&i; by id;
            proc sort data=fday&i; by id;
            data f&i;
                    merge ftrt&i fday&i;
                    by id;
            data fvalues;
                merge fvalues f&i;
                by id;
                if id ^= . then output;
                drop f;
            proc datasets;
                delete f&i ftrt&i fday&i;
       %end;
%MEND SIM;

  /* compute the observed F-statistic */
  proc glm data=one outstat=obsf;
     class sequence subject period trt day;
     model diff = sequence subject(sequence) period
         trt day;
     test h=sequence e=subject(sequence);
  run;

/* get observed F-values */
  data ftrt;
```

```
      set obsf;
       id = 1;
       if _type_ = 'SS3' and _source_ = 'TRT' then ftrt = f;
       if _type_ = 'SS3' and _source_ = 'TRT' then output;
       drop _type_ _source_ df prob _name_ ss;
  data fday;
      set obsf;
      id = 1;
      if _type_ = 'SS3' and _source_ = 'DAY' then fday = f;
      if _type_ = 'SS3' and _source_ = 'DAY' then output;
      drop _type_ _source_ df prob _name_ ss;
  proc sort data=ftrt; by id;
  proc sort data=fday; by id;
  data obsf;
      merge ftrt fday;
      by id;
      drop f id;
  %sim;
  proc iml;
      use fvalues var _all_;
      read all;
      use obsf var _all_;
      read all;
      ctrtrt = 0;
      ctrday = 0;
      do i = 1 to nrow(ftrtrs);
          if ftrtrs(i) >= ftrt then ctrtrt = ctrtrt + 1;
          if fdayrs(i) >= fday then ctrday = ctrday + 1;
      end;
      n = nrow(ftrtrs);
      ptrt = ctrtrt/n;
      pday = ctrday/n;
      reset center noname;
      mattrib ftrt format=8.4;
      mattrib fday format=8.4;
      mattrib ptrt format=8.4;
      mattrib pday format=8.4;
      mattrib n format=8.0;
      print 'Results of Randomization Test';
      print 'Observed Test Statistic (F-value) for
          Treatment: ' ftrt;
      print 'p-value Based On Resampled Distribution: '
          ptrt;
      print '    ';
      print 'Observed Test Statistic (F-value) for Day: '
          fday;
      print 'p-value Based On Resampled Distribution:    '
          pday;
      print 'Number of Resampling Simulations:          ' n;
      run;
      proc univariate data=fvalues normal;
              title1 'Summary Statistics for F-values';
              var ftrtrs fdayrs;
      run;
```

```
/***********************************************************
 *
 *   SAS Procedure to: RESAMPLE IN AN ANCOVA
 *
 ***********************************************************/
dm 'clear log';
dm 'clear output';

    /****************************************************
       Resampling Macro for ANCOVA
     ****************************************************/
    %MACRO RESAMPLE;
        proc iml;
        use resdata var _all_;
        read all;
        n = nrow(e);
        seed1 = time();
        u = ranuni(J(n, 1, seed1));
        i = int(n*u + j(n,1,1));
        newe = e(|i,|);
        newy = yhat + newe;
        create newdata var {trt newy rankpre};
        append var {trt newy rankpre};
        proc glm data=newdata outstat=outstat noprint;
                class trt;
                model newy = rankpre trt;
        run;
    %MEND STAT;

    %MACRO SIM;
        %do i = 1 %to 1000;
                %if &i = 250 %then dm 'clear log';
                %if &i = 500 %then dm 'clear log';
                %if &i = 750 %then dm 'clear log';
                %let id = &i;
                %resample;
                data f&i;
                    set outstat;
                    id = &i;
                    resampf = f;
                    if _type_ = 'SS3' and _source_ =
                'TRT' then output;
                    drop _type_ _source_ df prob _name_ ss;
                data fvalues;
                    merge fvalues f&i;
                    by id;
                    if id ^= . then output;
                    drop f;
                proc datasets;
                    delete f&i;
            %end;
    %MEND    SIM;
```

```
data one;
infile harass;
input trt pretest posttest;

/**************************************************
    do ANCOVA on ranked transformed data
 *************************************************/
    proc rank out=one;
            var pretest posttest;
            ranks rankpre rankpost;
    proc glm data=one outstat=obsf;
            class trt;
            model rankpost = rankpre trt;
            output out=resdata residual=e  predicted=yhat;
    run;
    data obsf;
            set obsf;
            if _type_ = 'SS3' and _source_ = 'TRT'
             then output;
            drop _type_ _source_ df prob _name_ ss;
    %sim;
    proc iml;
    use fvalues var _all_;
      read all;
      use obsf var _all_;
      read all;
      flt = choose(resampf < f, resampf, .);
      fgt = setdif(resampf, flt)`;
      n = nrow(resampf);
      p = nrow(fgt)/n;
      reset center noname;
      mattrib f format=8.4;
      mattrib p format=8.4;
      mattrib n format=8.0;
      print 'Results of Randomization Test';
      print 'Observed Test Statistic (F-value):        ' f;
      print 'p-value Based On Resampled Distribution: ' p;
      print 'Number of Resampling Simulations:        ' n;
    run;
    data fvalues;
      set fvalues;
      file fvalues;
      put resampf;
    proc univariate data=fvalues normal;
      title1 'Univariate Summary Statistics for Resampled
          F-values';
      var resampf;
    run;
```

REFERENCES

Altman, D.G. and Dore, C.J., Randomisation and baseline comparisons in clinical trials, *Lancet*, 335, 149, 1990.

Anastasi, A., *Psychological Testing*, Macmillian, New York, 1982.

Asnis, G.M., Lemus, C.Z., and Halbreich, U., The desipramine cortisol test - a selective noradrenergic challenge (relationship to other cortisol tests in depressives and normals), *Psychopharmacol. Bull.*, 22, 571, 1986.

Atiqullah, M., The robustness of the covariance analysis of a one-way classification, *Biometrika*, 51, 365, 1964.

Austin, H.A., III, Muenz, L.R., Joyce, K.M., Antonovych, T.A., Kullick, M.E., Klippel, J.H., Decker, J.L., and Balow, J.W., Prognostic factors in lupus nephritis, *Am. J. of Med.*, 75, 382, 1983.

Berry, D.A., Logarithmic transformations in ANOVA, *Biometrics*, 43, 439, 1987.

Berry, D.A., Basic principles in designing and analyzing clinical studies, in *Statistical Methodology in the Pharmaceutical Sciences*, Berry, D.A., Ed., Marcel Dekker, New York, 1990.

Birch, J. and Myers, R.H., Robust analysis of covariance, *Biometrics*, 38, 699, 1982.

Blair, R.C. and Higgins, J.J., Comparison of the power of the paired samples t-test to that of Wilcoxon's signed-ranks test under various population shapes, *Psychol. Bull.*, 97, 119, 1985.

Blom, G., *Statistical Estimates and Transformed Beta Variables*, John Wiley & Sons, New York, 1958.

Bonate, D.L. and Jessell, J.C., The effects of educational intervention on perceptions of sexual harassment, *Sex Roles*, 35, 751, 1996.

Bonett, D.G., On post-hoc blocking, *Educ. and Psychol. Meas.*, 42, 35, 1982.

Bowles, S.K., Reeves, R.A., Cardozo, L., and Edwards, D.J., Evaluation of the pharmacokinetic and pharmacodynamic interaction between quinidine and nifedipine, *J. of Clin. Pharmacol.* 33, 727, 1993.

Box, G.E.P. and Muller, M.E., A note on the generation of random normal variates, *Ann. Math. Stat.*, 29, 160, 1958.

Bradley, E.L. and Blackwood, L.G., Comparing paired data: a simultaneous test for means and variances, *Am. Stat.*, 43, 234, 1989.

Brogan, D.R. and Kutner, M.H., Comparative analyses of pretest-posttest research designs, *Am. Stat*, 34, 229, 1980.

Brouwers, P. and Mohr, E., A metric for the evaluation of change in clinical trials, *Clin. Neuropharmacol.*, 12, 129, 1989.

Burdick, W.P., Ben-David, M.F., Swisher, L., Becher, J., Magee, D., McNamara, R., and Zwanger, M., Reliability of performance-based clinical skill assessment of emergency room medicine residents, *Acad. Emerg. Med.*, 3, 1119, 1996.

Carmines, E.G. and Zeller, R.A., *Reliability and Validity*, Sage Publications, Newbury Park, CA, 1979.

Chen, S. and Cox, C., Use of baseline data for estimation of treatment effects in the presence of regression to the mean, *Biometrics*, 48, 593, 1992.

Chesher, A., Non-normal variation and regression toward the mean, *Stat. Meth. Med. Res.*, 6, 147, 1997.

Chuang-Stein, C., The *regression* fallacy, *Drug Inf. J.*, 27, 1213, 1993.

Chuang-Stein, C. and Tong, D.M., The impact and implication of regression to the mean on the design and analysis of medical investigations, *Stat. Meth. in Med. Res.*, 6, 115, 1997.

Cochran, W.G., Analysis of covariance: its nature and uses, *Biometrics*, 13, 261, 1957.

Cohen, J., *Statistical Power Analysis for the Behavioral Sciences*, Lawrence Erlbaum Associates, Hillsdale, NJ, 1988.

Conover, W.J. and Iman, R.L., Rank transformations as a bridge between parametric and nonparametric statistics (with discussion), *Am. Stat.*, 35, 124, 1981.

Conover, W.J. and Iman, R.L., Analysis of covariance using the rank transformation, *Biometrics*, 38, 715-724, 1982.

Crager, M.R., Analysis of covariance in parallel-group clinical trials with pretreatment baselines, *Biometrics*, 43, 895, 1987.

Cronbach, L.J. and Furby, L., How should we measure change-or should we?, *Psychol. Bull.*, 74, 68, 1970.

D'Agostino, R.B., Belanger, A., and D'Agostino, R.B., Jr., A suggestion for using powerful and informative tests of normality, *Am. Stat.*, 44, 316, 1990.

Dalton, S. and Overall, J.E., Nonrandom assignment in ANCOVA: the alternate ranks design, *J. Exp. Ed.*, 46, 58, 1977.

Davidian, M. and Giltinan, D.M., *Nonlinear Models for Repeated Measures Data*, Chapman & Hall, Boca Raton, FL, 1995.

Davis, C.E., The effect of regression to the mean in epidemiologic and clinical studies, *Am. J. of Epidemiol.*, 104, 493, 1976.

Dawson, J.D., Comparing treatment groups on the basis of slopes, area under the curves, and other summary measures, *Drug Inf. J.*, 28, 723, 1994.

DeGracie, J.S. and Fuller, W.A., Estimation slope and analysis of covariance when the concomitant variable is measured with error, *J. of the Am. Stat. Assoc.*, 67, 930, 1972.

de Mey, C. and Erb, K.A., Usefulness, usability, and quality criteria for noninvasive methods in cardiovascular pharmaccology, *J. Clin. Pharmacol.*, 27, 11S, 1997.

Diggle, P.J., Liang, K.-Y., and Zeger, S.L., *Analysis of Longitudinal Data*, Clarendon Press, Oxford, 1995.

Donahue, R.M.J., A summary statistic for measuring change from baseline, *J. of Biopharm. Stat.*, 7, 287, 1997. Marcell Dekker, Inc., New York.

Enas, G.G., Enas, N.H., Spradlin, C.T., Wilson, M.G., and Wiltse, C.G., Baseline comparability in clinical trials: prevention of "poststudy anxiety," *Drug Inf. J.*, 24, 541, 1990.

Feldt, L.S., A comparison of the precision of three experimental designs employing a concomitant variable, *Psychometrika*, 23, 335, 1958.

Ferguson, G.A. and Takane, Y., *Statistical Analysis in Psychology and Education*, McGraw-Hill, New York, 1989.

Fries, J.F., Porta, K., and Liang, M.H., Marginal benefit of renal biopsy in systemic lupus erythematosus, *Arch. of Int. Med.*, 138, 1386, 1978.

Frison, L.J. and Pocock, S.J., Repeated measures in clinical trials: analysis using mean summary statistics and its implication for design, *Stat. in Med.*, 11, 1685, 1992.

Furby, L., Interpreting regression toward the mean in developmental research, *Dev. Psychol.*, 8, 172, 1973.

Gail, M.H., Tan, W.Y., and Piantadosi, S., Tests for no treatment effect in randomized clinical trials, *Biometrika*, 75, 57, 1988.

Galton, F., Regression towards mediocrity in hereditary stature, *J. of Anthropol. Inst.*, 15, 246, 1885.

George, V., Johnson, W.D., Shahane, A., and Nick, T.G., Testing for treatment effect in the presence of regression towards the mean, *Biometrics*, 53, 49, 1997.

Ghiselli, E.E., *Theory of Psychological Measurement*, McGraw-Hill, New York, 1964.

Ghiselli, E.E., Campbell, J.P., and Zeddeck, S., *Measurement Theory for the Behavioral Sciences*, W.H. Freeman, San Francisco, 1981.

Gill, J.L., Heterogeneity of variance in randomized block experiments, *J. of Anim. Sci.*, 59, 1339, 1984.

Glass, G.V., Peckham, P.D., and Sanders, J.R., Consequences of failure to meet assumptions underlying the fixed effects analyses of variance and covariance, *Rev. Ed. Res.*, 42, 237, 1972.

Good, P., *Permutation Tests: A Practical Guide to Resampling Methods for Testing Hypotheses*, Springer-Verlag, New York, 1994.

Graham, J.R., *The MMPI: A Practical Guide*, 2nd ed., Oxford University Press, New York, 1987.

Grambsch, P.M., Simple robust tests for scale differences in paired data, *Biometrika*, 81, 359, 1994.

Guilford, J.P., *Psychometric Methods*, McGraw-Hill, New York, 1954.

Hamilton, B.L., A Monte Carlo test for robustness of parametric and nonparametric analysis of covariance against unequal regression slopes, *J of the Am. Stat. Assoc.*, 71, 864, 1976.

Huck, S.W. and McLean, R.A., Using a repeated measures ANOVA to analyze the data from a pretest-posttest design: a potentially confusing task, *Psychol. Bull.*, 82, 4, 1975.

Jaccard, J., *Interaction Effects in Factorial Analysis of Variance*, Sage Publications, Thousand Oaks, CA, 1997.

Jaccard, J., Turriso, R., and Wan, C.K., *Interaction Effects in Multiple Regression*, Sage Publications, Newbury Park, CA, 1990.

James, K.E., Regression toward the mean in uncontrolled clinical studies, *Biometrics*, 29, 121, 1973.

Kaiser, L., Adjusting for baseline: change or percentage change?, *Stat. in Med.*, 8, 1183, 1989.

Kirk, R.E., *Experimental Design: Procedures for the Behavioral Sciences*, Brooks/Cole, Belmont, CA, 1982.

Knoke, J.D., Nonparametric analysis of covariance for comparing change in randomized studies with baseline variables subject to error, *Biometrics*, 47, 523, 1991.

Kuhn, A. and DeMasi, R.A., Empirical power of distribution-free tests of incomplete longitudinal data with applications to AIDS clinical trials, *J. of Biopharm. Stat.*, 9, 401, 1999.

Labouvie, E.W., The concept of change and regression toward the mean, *Psychol. Bull.*, 92, 251, 1982.

Lacey, L.F., O'Keene, O.N., Pritchard, J.F., and Bye, A., Common noncompartmental pharmacokinetic variables: are they normally or log-normally distributed?, *J. of Biopharm. Stat.*, 7, 171, 1997.

Laird, N., Further comparative analyses of pretest-posttest research designs, *Am. Stat.*, 37, 329, 1983.

Lana, R.E., Pretest sensitization, in *Artifact in Behavioral Research,* Rosenthal, R. and Rosnow, R.L, Eds., Academic Press, London, 1969.

Levine, P.H., An acute effect of cigarette smoking on platelet function, *Circulation*, 48, 618, 1973.

Lin, H.M. and Hughes, M.D., Adjusting for regression toward the mean when variables are normally distributed, *Stat. Meth. in Med. Res.,* 6, 129, 1997.

Lipsey, M.W., *Design Sensitivity: Statistical Power for Experimental Research*, Sage Publications, Newbury Park, CA, 1990.

Littell, R.C., Milliken, G.A., Stroup, W.W., and Wolfinger, R.D., *SAS System for Mixed Models,* SAS Institute, Inc., Cary, NC, 1996.

Lord, F.M., Large-sample covariance analysis when the control variable is fallible, *J of the Am. Stat. Assoc.*, 55, 307, 1955.

Lord, F.M., Elementary models for measuring change, in *Problems in Measuring Change,* Harris, C.W., Ed., University of Wisconsin Press, Madison WI, 1963.

Lord, F.M. and Novick, M.R., *Statistical Theories of Mental Test Scores*, Addison-Wesley, Reading, MA, 1968.

Ludbrook, J. and Dudley, H., Why permutation tests are superior to t and F tests in biomedical literature, *Am. Stat.*, 52, 127, 1998.

Maxwell, S.E., Delaney, H.D., and Dill, C.A., Another look at ANCOVA versus blocking, *Psychol. Bull.*, 95, 136, 1984.

McCullogh, C.E., Tests for equality of variances with paired data, *Commun. in Stat. Ser. − Theory and Meth.*, 16, 1377, 1987.

McDonald, C.J., Mazzuca, S.A., and McCabe, G.P., How much placebo 'effect' is really statistical regression?, *Stat. in Med.*, 2, 417, 1983.

McNeil, D., On graphing paired data, *Am. Stat.*, 46, 307, 1992.

Mee, R.W. and Chua, T.C., Regression toward the mean and the paired sample t-test, *Am. Stat.*, 45, 39, 1991.

Menegazzi, J.J., Davis, E.A., Sucov, A.N., Paris, and P.N., Reliability of the Glasgow Coma Scale when used by emergency physicians and paramedics, *J. Trauma*, 34, 46, 1993.

Mooney, C.Z., *Monte Carlo Simulation*, Sage Publications, Thousand Oaks, CA, 1997.

Morgan, W.A., A test for the significance of the difference between two variances in a sample from a normal bivariate population, *Biometrika*, 31, 9, 1939.

Moye, L.A., Davis, B.R., Sacks, F., Cole, T., Brown, L., and Hawkins, C.M., Decision rules for predicting future lipid values in screening for a cholesterol reduction clinical trial, *Controlled Clin. Trials*, 17, 536, 1996.

Nesselroade, J.R., Stigler, S.M., and Baltes, P.B., Regression toward the mean and the study of change, *Psychol. Bull.*, 88, 622, 1980.

Neter, J., Kutner, M.H., Nachtsheim, C.J., and Wasserman, W., *Applied Linear Statistical Models*, Irwin, Chicago, 1996.

Noreen, E.W., *Computer-Intensive Methods for Testing Hypotheses: An Introduction*, John Wiley & Sons, New York, 1989.

Olejnik, S.F. and Algina, J., Parametric ANCOVA and the rank transform ANCOVA when the data are conditionally non-normal and heteroscedastic, *J. Educ. Stat.*, 9, 129, 1984.

Overall, J.E., Letter to the editor: the use of inadequate corrections for baseline imbalance remains a serious problem, *J. Biopharm. Stat.*, 3, 271, 1993.

Overall, J.E. and Ashby, B., Baseline corrections in experimental and quasi-experimental clinical trials, *Neuropsychopharmacol.*, 4, 273, 1991.

Peterson, R.G., *Design and Analysis of Experiments*, Marcel Dekker, New York, 1985.

Pittman, E.J.G., A note on normal correlation, *Biometrika*, 31, 9, 1939.

Pocock, S.J., *Clinical Trials: A Practical Approach*, John Wiley & Sons, New York, 1983.

Pratt, C.M., Ruberg, S., Morganroth, J., McNutt, B., Woodward, J., Harris, S., Ruskin, J., and Moye, L., Dose-response relation between terfenadine (Seldane) and the QTc interval on the scalar electrocardiogram: distinguishing a drug effect from spontaneous variability, *Am. Heart J.*, 131, 472, 1996.

Puri, M.L. and Sen, P.K., Analysis of covariance based on general rank scores, *Ann. of Math. Stat.*, 40, 610, 1969.

Quade, D., Rank analysis of covariance, *J. Am. Stat. Assoc.*, 62, 1187, 1967.

Raboud, J.M., Montaner, J.S.G., Rae, S., Conway, B., Singer, J., and Scheter, M.T., Issues in the design and trials of therapies for HIV infected individuals with plasma RNA level as an outcome, *J. of Infect. Dis.*, 175, 576, 1996.

Rosenbaum, P.R., Exploratory plots for paired data, *Am. Stat.*, 43, 108, 1989.

Sandvik, L. and Olsson, B., A nearly distribution-free test for comparing dispersion in paired samples, *Biometrika*, 69, 484, 1982.

SAS Institute, *SAS/STAT Users Guide*, Version 6, SAS Institute, Cary, NC, 1990.

Seaman, S.L., Algina, J., and Olejnik, S.F., Type I error probabilities and power of the rank and parametric ANCOVA procedures, *J. of Educ. Stat.*, 10, 345, 1985.

Seber, G.A.F., *Linear Regression Analysis*, John Wiley & Sons, New York, 1977.

Senn, S., Testing for baseline differences in clinical trials, *Stat. in Med.*, 13, 1715, 1994.

Senn, S. and Brown, R., Maximum likelihood estimation of treatment effects for samples subject to regression to the mean, *Commun. Stat. Ser. − Theory and Methods*, 18, 3389, 1989.

Senn, S.J. and Brown, R.A., Estimating treatment effects in clinical trials subject to regression to the mean, *Biometrics*, 51, 555, 1985.

Shapiro, S.S. and Wilk, M.B., An analysis of variance test for normality (complete samples), *Biometrika*, 52, 591, 1965.

Shepard, D.S., Reliability of blood pressure measurements: implications for designing and evaluating programs to control hypertension, *J. of Chronic Dis.*, 34, 191, 1981.

Snow, W.G., Tierney, M.C., Zorzitto, M.L., Fisher, R.H., and Reid, D.W., WAIS-R test-retest reliability in a normal elderly sample, *J. of Clinical Exp. Neuropsychol.*, 11, 423, 1989.

Solomon, R.L., An extension of control group design, *Psychol. Bull.*, 46, 137, 1949.

Stigler, S.M., Regression toward the mean, historically considered, *Stat. Meth. in Med. Res.*, 6, 103, 1997.

Suissa, S., Levinton, C., and Esdaile, J.M., Modeling percentage change: a potential linear mirage, *J. of Clin. Epidemiol.*, 42, 843, 1989.

Tornqvist, L., Vartia, P., and Vartia, Y.O., How should relative change be measured?, *Am. Stat.*, 39, 43, 1985.

Tufte, E.R., *The Visual Display of Quantitative Information*, Graphics Press, Cheshire, CT, 1983.

Tukey, J.W., The future of data analysis, *Ann. of Math. Stat.*, 33, 22, 1962.

van der Ent, C.K. and Mulder, P., Improvement in tidal breathing pattern analysis in children with asthma by on-line automatic data processsing, *Eur. Respiratory J.*, 9, 1306, 1996.

Wainer, H., Adjusting for differential rates: Lord's paradox again, *Psychol. Bull.*, 109, 147, 1991.

West, S.M., Herd, J.A., Ballantyne, C.M., Pownall, H.J., Simpson, S., Gould, L., and Gotto, A.M., The Lipoprotein and Coronary Atherosclerosis Study (LCAS): design, methods, and baseline data of a trial of fluvastatin in patients without severe hypercholesteremia, *Controlled Clin. Trials*, 17, 550, 1996.

Whiting-O'Keefe, Q., Henke, J.E., Shearn, M.A., Hopper, J., Biava, C.G., and Epstein, W.V., The information content from renal biopsy in systemic lupus erythematosus, *Ann. of Int. Med.*, 96, 718, 1982.

Wilcox, R.R., Comparing the variances of dependent groups, *Psychometrika*, 54, 305, 1989.

Wilcoxon, F., Individual comparisons by ranking methods, *Biometrics Bull.*, 1, 80, 1945.

Wilder, J., Adrenalin and the law of initial values, *Exp. and Med. Surg.*, 15, 47, 1957.

Wildt, A.R. and Ahtola, O., *Analysis of Covariance*, Sage Publications, Beverly Hills CA, 1978.

Williams, R.H. and Zimmerman, D.W., Are simple gain scores obsolete?, *Appl. Psychol. Meas.*, 20, 59, 1996.

Wolfinger, R.D., An example of using mixed models and PROC MIXED for longitudinal data, *J. of Biopharm. Stat.*, 7, 481, 1997.

Yuan, C.-S., Foss, J.F., Osinski, J., Toledano, A., Roizen, M.F., and Moss, J., The safety and efficacy of oral methylnaltrexone in preventing morphine-induced delay in oral cecal-transit time, *Clin. Pharmacol. and Therap.*, 61, 467, 1997.

Zar, J.H., *Biostatistical Analysis*, 2nd ed., Prentice-Hall, Englewood Cliffs, NJ, 1984.

INDEX